CABI CONCISE

MEDICINAL AND AROMATIC PLANTS

Aromatic plants and spices are probably the most underestimated food worldwide. Although they have constantly accompanied humans in the course of evolution (co-evolution), they are only rarely found in official dietary recommendations. Yet they are not only aromatic, but extremely effective and can be used for everything from increasing well-being to treating various illnesses. They are, so to speak, the "tastiest medicine in the world". The essential oils contained in most spices, for example, not only give the spice its specific flavour, but also have various antibacterial, antiviral and anti-inflammatory properties.

There is a very long tradition of using herbs and spices for health. In traditional medicinal systems such as Ayurvedic, Chinese European Monastic and Persian medicine, spices have been used to treat various diseases for more than 3,000 years. Today, traditional medicinal systems are a topic of global importance. In many countries, a large proportion of the population relies heavily on traditional practitioners and medicinal plants to meet the need for basic medical care. This is because modern synthetic preparations are often only available to a limited extent and are usually very expensive. Medicinal and aromatic plants (MAPs), or herbal medicines, on the other hand, have often retained their popularity for historical and cultural reasons.

In recent decades, increasing attention has been paid to the phytochemical and pharmacological study of traditional medicinal plants in order to discover and develop new, safe and affordable medicines. These medicinal plants are not only derived from the well-known traditional medical systems, but also refer to the experiential knowledge of indigenous people and tribes in South America and Africa. For example, the root of devil's claw (*Harpagophytum procumbens*) was used extensively for healing purposes by the first inhabitants of South Africa.

Since human olfactory cells have direct access to the oldest regions of our brain, such as the limbic system and the hippocampus, where emotions, feelings and, above all, our memories are stored, scents have a direct effect on our well-being. It therefore seems worthwhile to consider how traditional knowledge about the health benefits of various medicinal and aroma plants can be re-evaluated using the techniques available today, and how the results can be utilized for the pharmaceutical and food sectors.

This series focuses on traditional medicine systems and covers different medicinal plant species or products. Each book will largely follow the same content structure, and will cover the history, efficacy, active ingredients, occurrence, genetics/breeding, cultivation and economic importance.

As the market demand for natural ingredients is increasing, this series aims to provide succinct and readable overviews of key medicinal and aromatic plant species used around the world. We welcome new proposals and suggestions.

Hartwig Schulz
hs.consulting.map@t-online.de

CABI is a trading name of CAB International

CABI
Nosworthy Way
Wallingford
Oxfordshire OX10 8DE
UK

CABI
200 Portland Street
Boston
MA 02114
USA

Tel: +44 (0)1491 832111
E-mail: info@cabi.org
Website: www.cabi.org

T: +1 (617)682-9015
E-mail: cabi-nao@cabi.org

A catalogue record for this book is available from the British Library, London, UK.

A catalogue record for this book is available from the British Library, London, UK.

Library of Congress Control Number: 202594444

ISBN-13: 9781836991199 (hardback)
9781836991205 (paperback)
9781836991212 (OA ePDF)
9781836991229 (OA ePub)

DOI: 10.1079/9781836991229.0000

Commissioning Editor: Rebecca Stubbs
Editorial Assistant: Emma McCann
Production Editor: Rosie Hayden

Typeset by Exeter Premedia Services Pvt Ltd, Chennai, India

Essential Oils Unveiled

Complex Compositions for Food, Cosmetics, and Medicine

Marek Bunse, Hartwig Schulz,
Cäcilia Brendieck-Worm, Rolf Daniels,
Sandra Graf-Schiller, Jörg Heilmann,
Dietmar R. Kammerer, Matthias F. Melzig,
Gertrud E. Morlock, Constanze Stiefel,
Florian C. Stintzing, Michael Wink,
Eliane Zimmermann

CABI

Contents

Authors

Brendieck-Worm, Cäcilia Phyto Fokus, Talstraße 59, 67700 Niederkirchen, Germany

Bunse, Marek Department of Science, WALA Heilmittel GmbH, 73087 Bad Boll, Germany

Daniels, Rolf Department of Pharmaceutical Technology, University of Tübingen, 72074 Tübingen, Germany

Graf-Schiller, Sandra SaluVet GmbH, 88339 Bad Waldsee, Germany

Heilmann, Jörg Department of Pharmaceutical Biology, University of Regensburg, 93053 Regensburg, Germany

Kammerer, Dietmar R. Department of Science, WALA Heilmittel GmbH, 73087 Bad Boll, Germany

Melzig, Matthias F. Institute of Pharmacy, Freie Universität Berlin, 14195 Berlin, Germany

Morlock, Gertrud E. Institute of Nutritional Science, Department of Food Science and TransMIT Center for Effect-Directed Analysis, Justus Liebig University Giessen, 35392 Giessen, Germany

Schulz, Hartwig Consulting & Project Management for Medicinal & Aromatic Plants, 14532 Stahnsdorf, Germany

Stiefel, Constanze Esslingen University of Applied Science, Science, Energy and Building Services, 73728 Esslingen, Germany

Stintzing, Florian C. Department of Science, WALA Heilmittel GmbH, 73087 Bad Boll, Germany

Wink, Michael Institute of Pharmacy and Molecular Biotechnology, Heidelberg University, 69120 Heidelberg, Germany

Zimmermann, Eliane Aromapraxis, www.aromapraxis.de, Ardaturrish More, P75 E427 Glengarriff, County Cork, Ireland

Preface

This multiple-author book is an initiative of experts working on mixtures of plant metabolites, which are natural complex substances, and brings together decades of scientific knowledge from different disciplines.

This work is aimed at all readers who want to get an overview of all fields in which essential oils have been playing a role in human history. Essential oils are still of high relevance and will be vital in the future. They are natural complex substances occurring in versatile mixtures by their very nature, i.e. they exist as distinct compositions borne by nature for specific evolutionary purposes. Unveiling this logic and discovering the secrets of this world of interconnected substances is worthwhile at a time when a 'one substance-one target' approach is widespread in the scientific and medicinal world.

Essential oils have been part of human culture for as long as humans have existed. Humans can profit from their broad potential for use in food, cosmetics, and human and veterinary medicines. Profound knowledge, as discussed in this monograph, helps to identify those potential fields of application where essential oils may have been overlooked so far. It also prevents hazard-based approaches and paves the way for a stronger risk-based evaluation of essential oils. Moreover, essential oils and other complex mixtures of secondary metabolites can be helpful in times of increasing antibiotic resistance and they can contribute to planetary health.

We would like to thank all the co-authors, who made this book possible through their commitment and enthusiasm. We are also grateful to our editors Hartwig Schulz and Marek Bunse, who took care of the chapters, converting them to a homogeneous layout while respecting the individual styles of individual authors. We are grateful to CABI, especially Rebecca Stubbs, who paved the way for this multiple-author work to come together in one book in a most cooperative and straightforward way.

This book was financed by the Stiftung Integrative Medizin & Pharmazie, Stuttgart/Germany (www.stintmed.de), which supports initiatives dedicated to natural complex substances.

We hope that this book finds an interested readership and widespread distribution as an introduction to the interesting world of essential oils.

Dietmar R. Kammerer

Florian C. Stintzing

Essential Oils as Complex Mixtures – Fundamentals and History

Marek Bunse, Dietmar R. Kammerer and Florian C. Stintzing*

1.1 History of Phytotherapy

Aromatic and medicinal plants have always been used in human civilization. Whether for food as herbs, spices or flavouring, or for medicinal applications to treat several conditions like inflammation or microbial infections (Freitas and Cattelan, 2018). The strong connection between natural complex substances (NCS) and humans, including their common evolution, underscores the potential of nature and natural substances for health and well-being.

In particular, phytotherapy and the use of medicinal plant extracts represent the most recognized branches of natural medicine in Europe today. The roots of European phytotherapy can be tracked back to antiquity with books such as the *De Materia Medica* written by Dioscorides (Staub *et al.*, 2016). This book was one of the first pharmacognostic guides, which describes plants, animals and the compounds derived from them, like fatty acids or essential oils (EOs).

In the period 500–1500 AD, the so-called Middle Ages, the cultivation and application of medicinal plants was carried out on a large scale by monasteries. New extracts and formulations were developed from different natural substances, which played a crucial role in the treatment of health complaints. By the 13th century, monks and pharmacists became increasingly important in the preparation and application of herbal remedies (Dufault *et al.*, 2001).

In the 14th century, the Black Death, the bubonic plague pandemic, revealed the limits of medieval healthcare knowledge. Approximately 25% of

*Corresponding author: Florian.stintzing@wala.de

the European population was decimated by the pandemic, which led to social restrictions and setbacks in nearly all areas, including phytotherapy, nursing and medicine. Poor living conditions, owing to famine and inadequate sanitation, combined with agricultural failures, led to further health complications and the spread of diseases.

1.1.1 Renaissance and the expansion of herbal knowledge

The late 15th century marked the beginning of rapid research and knowledge dissemination, which had a profound impact on phytotherapy. The discovery of the New World and the establishment of sea routes to India provided access to previously unknown plants and a variety of new opportunities. Exotic plants such as cacao, chili pepper and sunflower, with their unique ingredients and multiple biological properties, expanded the pool of possible new applications, including the treatment of diseases. With the development of printing, detailed illustrations and descriptions of medicinal plants became accessible to a broad audience and were no longer confined to handwritten manuscripts. Notable herbal books from this period were Otto Brunfels's *Herbarium Vivae Eicones* (1530), Leonhart Fuchs's *De Historia Stirpium* (1542), John Gerard's *Herbarium* (1597), Nicholas Culpeper's *Der englische Arzt* (1649) and John Parkinson's *Theatrum Botanicum* (1669), which played an important role in spreading botanical and medicinal knowledge (Makarska-Białokoz, 2020).

1.1.2 Paracelsus: A paradigm shift in phytotherapy

Paracelsus (Philippus Theophrastus Aureolus Bombastus von Hohenheim, 1493–1541) was also an important personality who revolutionized herbal medicine during the Renaissance. He emphasized the importance of observation, experimentation and a deeper understanding of the chemical properties of plants and questioned the traditional Galenic theory of humours. His famous statement was: 'Poison is in everything, and no thing is without poison; the dose makes it either a poison or a remedy', underscoring the importance of dose-dependent effects and the potential toxicity of every substance or compound (Holzinger, 2013). This groundbreaking insight was the beginning of modern toxicology and pharmacology, including research in phytochemistry and pharmacognosy (Michaleas *et al.*, 2021).

1.1.3 The rise of modern chemistry and its impact on phytotherapy

The 18th and 19th centuries became the centuries of chemistry, in which natural science developed a more chemical profile. Research focused on compound characterization and identification of individual bioactive substances occurring in medicinal plants to explain their therapeutic effects. The holistic approach to herbal medicine of the past several centuries has changed to an in-depth examination of individual active components that appear to be specific to the healing power of plants. New methods have been developed to purify and isolate these specific compounds with the aim of producing standardized bioactive extracts that can be easily dosed. The isolation of morphine from the opium poppy by pharmacist Friedrich Wilhelm Adam Sertürner in 1806 was a pivotal moment in this chemical revolution (Sertürner, 1806). This discovery led to more studies, which resulted in the identification of numerous other highly bioactive natural compounds such as strychnine, quinine, caffeine, salicin, cocaine and digitalin (Barnes, 2007).

The development and use of chromatography, microscopy and other advanced analytical methods also expanded scientific knowledge of medicinal plants. Thereafter, scientists were able to analyse the chemical composition of plant extracts, to assess their bioactivity profiles and the pharmacological and toxicological risks associated with their use, and to establish guidelines on how to ensure the quality of plant-derived products and extracts with standardized methods. This period was the beginning of a new era in pharmaceutical research where plants served as a rich source of novel medicines. However, this swiftness towards the isolation and syn-thesis of active metabolites and derived components also marked the decline of using the entire plant extract. In the 1930s, synthetic pharmaceutical ingredients, which were often inspired by plant molecules, dominated the pharmaceutical market (Ferreira *et al.*, 2014; Jamshidi-Kia *et al.*, 2018; Makarska-Białokoz, 2020).

1.1.4   The reemergence of rational phytotherapy

Drug discovery has revolutionized the isolation of plant compounds and the development of synthetic substances, but in contrast to this trend, a movement has emerged that is explicitly committed to the holistic approach to herbal medicine. This movement is known as rational phytotherapy, which bridges the gap between traditional knowledge and experiences from past centuries and modern science.

Rational phytotherapy emphasizes:

- **scientific validation:** rigorous research to confirm the efficacy and safety of traditional herbal remedies;
- **standardized cultivation and processing:** implementation of guide-lines like the 'Good agricultural and collection practice' (WHO, 2003) and the 'Guideline on quality of herbal medicinal products / traditional herbal medicinal products' (EMA/CPMP/QWP/2819/00; EMA, undated) to ensure the quality and consistency of herbal products (Sahoo *et al.*, 2010; Fürst and Zündorf, 2015); and

- **understanding synergistic effects:** recognizing that the complex mixture of compounds in plants often work together to produce therapeutic benefits, which may be lost when isolating individual constituents.

This scientific approach has resulted in the creation of standardized herbal remedies with enhanced quality assurance and clinical evidence supporting their application.

1.1.5   The enduring potential of medicinal plants

Medicinal plants are still of great relevance in today's world. In particular, the search for effective, affordable and clinically proven medicines is more important than ever. The knowledge about the traditional use of plants as medicines, coupled with the chemical diversity they offer, makes them a promising source for drug development.

Plants have developed numerous mechanisms to counteract various biotic and abiotic stressors, which has led to the creation of a wide set of bioactive compounds that may be used for therapeutic purposes (Wink, 2003; Sakkas and Papadopoulou, 2017). These include:

- **anti-inflammatory agents:** compounds that reduce inflammation, which is a key factor in many chronic diseases. For example, curcumin from turmeric (*Curcuma longa* L.) roots (Gonfa *et al.*, 2023);
- **antibacterial, antifungal and antiviral agents:** substances that may combat infectious diseases, like resveratrol, which shows promising antibacterial activities, e.g. against *Staphylococcus aureus (Stan et al., 2021)*;
- **anticancer agents:** compounds that inhibit the growth and spread of cancer cells, like lectins and viscotoxins from mistletoe (*Viscum album* L.; Yosri *et al.*, 2024); and
- **antiparasitic agents:** substances, e.g. polyphenols such as tannins or terpenoids like geraniol, which attack and kill parasites (Ranasinghe *et al.*, 2023).

Despite their vast potential, the pharmacological and clinical effects of medicinal plants have not been fully explored, but there is a growing interest in a deeper understanding of their composition and biological properties, as well as a need for more profound scientific research to:

- **validate traditional uses:** confirm their efficacy and safety using state-of-the-art scientific methods;
- **uncover novel therapeutic applications:** explore the potential of little-known plants and their bioactive constituents; and
- **elucidate modes of action:** understand how plant-derived substances interact with the human body on a molecular level.

Thus, by integrating the ancient knowledge in the use of medicinal plants with this contemporary scientific approach, the effective use of medicinal plant resources and new ways of treating various diseases may be found. This means that plant-derived medicines can be used together with synthetic and/or isolated agents, which are generally assumed to act on a single target.

1.2   The Intricate World of Medicinal Plant Complexity

The unique therapeutic potential of medicinal plants can be traced back to their diverse natural complexity of compounds. Nature does not have one single 'magic bullet' to interact with its environment; the effectiveness of natural complex substances is based on the complex interaction of numerous ingredients that form a symphony of biological activities.

1.2.1   Multi-component mixtures: A symphony of molecules

Medicinal plants are NCS systems, which are composed of a diverse array of bioactive compounds (Wink, 2003, 2024; Xie *et al.*, 2016). These include:

- **secondary metabolites:** alkaloids, terpenoids, phenolic compounds and other specialized molecules produced by plants for the defence against biotic and abiotic stress, communication and other ecological roles (Fig. 1.1); and
- **non-coding RNAs:** emerging evidence suggests that small RNA species, such as microRNAs, also contribute to the therapeutic effects of plants.

1.2.2   The power of synergy and the limits of isolation

Interestingly, isolated plant substances often appear to have a weaker therapeutic effect than in their natural form as plant extract with all the other compounds in the plant matrix. This highlights a crucial aspect of NCS: The whole is greater than the sum of its parts.

Several factors contribute to this phenomenon (Chou, 2006; Hemaiswarya *et al.*, 2008):

- **Synergy:** Individual compounds of a complex mixture may enhance the activity of each other, resulting in a combined effect greater than the sum of their individual contributions.
- **Additive effects:** Constituents may complement each other's actions, targeting multiple pathways or mechanisms involved in a particular disease.
- **Antagonism:** Some compounds may mitigate potential adverse side effects of others, contributing to a more balanced therapeutic profile, allowing a safer application.

The synergy, additivity and antagonism in the interaction underline the limits of a reductionist approach to herbal medicine. Single-substance isolation and investigation may never represent the complex biological activity of crude extracts of medicinal plants in their wholeness.

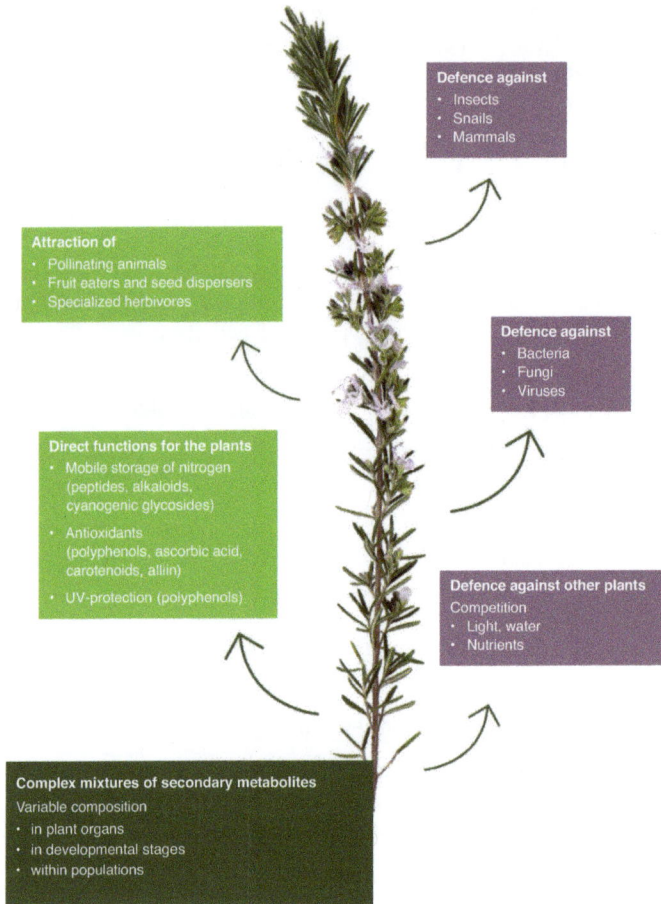

Defence against
- Insects
- Snails
- Mammals

Attraction of
- Pollinating animals
- Fruit eaters and seed dispersers
- Specialized herbivores

Defence against
- Bacteria
- Fungi
- Viruses

Direct functions for the plants
- Mobile storage of nitrogen (peptides, alkaloids, cyanogenic glycosides)
- Antioxidants (polyphenols, ascorbic acid, carotenoids, alliin)
- UV-protection (polyphenols)

Defence against other plants
Competition
- Light, water
- Nutrients

Complex mixtures of secondary metabolites
Variable composition
- in plant organs
- in developmental stages
- within populations

Fig. 1.1. Natural complex substances and their role in the plant kingdom. (According to Wink, 2024. Author's own image.)

Case in point: Individual vs combined effects

When tested in isolated form, neither (±)-camphor nor α-terpineol exhibited any significant anticancer activity. Linalyl acetate showed only marginal effects. Combining terpineol with linalyl acetate led to a moderate increase in activity, reducing cancer cell proliferation by 33% and 45%, respectively, at a concentration of 10^{-3} M each. Remarkably, when all three compounds were used in combination in their naturally occurring proportions, a significant synergistic effect emerged. This combination reduced proliferation in two human colon cancer cell lines (HCT-116 p53$^{+/+}$ and p53$^{-/-}$) by 50% and 64%, respectively, at the same concentration (Itani *et al.*, 2008).

1.2.3   Embracing complexity for holistic therapies

Although the multi-component nature of medicinal plants is a challenge for analysis and scientific evaluation, it also defines their therapeutic value. Identifying this complexity is crucial for:

- developing effective phytotherapy;
- understanding traditional knowledge; and
- discovering novel therapeutic strategies.

Thus, by acknowledging the complexity of medicinal plants and applying sophisticated analytical methods, it is possible to unleash the therapeutic potential of plants and create more effective and less harmful treatments for numerous diseases. This could also help to minimize some adverse effects of drugs which are made from single chemical compounds. Furthermore, there is still much to be discovered about the interactions between synthetic and natural medicines, which may be beneficial for patient needs.

1.2.4   Multi-target action: Embracing the versatility of NCS

In contrast to single compounds, which usually have only one specific biological target, NCS, with their diverse compositions of minor and main compounds, can target multiple molecular sites, also known as broad-spectrum activities. Each compound in this composition has its own molecular structure with unique functional groups and chemical properties that can interact with different biological targets, which may provide several benefits (Nahrstedt and Butterweck, 2010; Schwabl *et al.*, 2013):

- **Increased efficacy:** By targeting multiple pathways involved in a disease, NCS can potentially achieve greater therapeutic effects than single-compound drugs.
- **Reduced resistance:** The chance of developing resistance to a multi-target therapeutic agent is quite low as compared to the single-target drugs as several simultaneous mutations are needed to escape the therapeutic actions.
- **Fewer side effects:** The lower concentrations of individual constituents in NCS, combined with their potential synergistic interactions, may contribute to a more favourable safety profile and fewer off-target effects.

There are numerous studies examining these interactions between NCS and metabolic pathways and how they influence biological processes (Morlock, 2021; Schreiner *et al.*, 2021). For example, studies have shown that NCS can:

- **modulate drug absorption and metabolism:** the bioavailability and efficacy of other medications taken concurrently may affect the absorption and metabolism of the drug (Karalis *et al.*, 2008); and
- **influence taste and smell perception:** this can have an impact on palatability and sensory experiences associated with food and medication.

Complex diseases where various interconnected pathways are involved may need medicinal agents with multiple target activities such as NCS. Further research in these fields is crucial to understand the complex mechanisms and their broad-spectrum effects in order to effectively harness their therapeutic potential.

Case in point: Lavender essential oil

Lavender (*Lavandula angustifolia* Mill.; Fig. 1.2) essential oil with its multi-component mixture excellently illustrates this concept. Linalool and linalyl acetate are often referred to as primary constituents of the EO, while more than 100 other compounds contribute to its characteristic aroma and therapeutic properties. Research suggests that the calming and sleep-promoting effects of lavender oil arise from the synergistic interactions of multiple components, including linalyl acetate, terpinen-4-ol, the main constituent linalool and even trace components (Cavanagh and Wilkinson, 2002; Lillehei and Halcon, 2014; Chen *et al.*, 2024).

Fig. 1.2. Lavender (*Lavandula angustifolia* Mill.) flowers. (Author's own image, courtesy of WALA Heilmittel GmbH.)

1.2.5   The allure of simplicity: Mono-substances vs NCS

Although NCS appear to have great potential for various therapies, the use of isolated mono-substances is increasingly preferred in health care and pharmacy. These preferences are due to several reasons:

• simplified quality control;
• standardization and application;
• straightforward preclinical and clinical studies; and
• easier identification of targets and mode of action.

Furthermore, ensuring the authenticity of NCS, such as EOs, poses a significant challenge. Sophisticated analytical techniques are often required to detect adulteration or synthetic additives, further increasing the complexity of their development and regulation (Do *et al.*, 2015).

However, this pursuit of simplicity comes at a cost. By focusing on mono-substances, the potential synergistic benefits of NCS may be overlooked. Recent research has increasingly highlighted the existence of synergistic interactions of natural NCS, demonstrating enhanced efficacy and potentially novel therapeutic applications (Bunse *et al.*, 2022).

Case in point: Quenching effect on skin sensitization

Adversarial interactions or quenching can be used to minimize skin sensitization by some EO components. One of the common EO components, cinnamaldehyde, is known to have sensitizing effects on the skin of some people. Research has also established that (+)-limonene can sometimes inhibit skin sensitization caused by cinnamaldehyde. This quenching effect was even more obvious when (+)-limonene was used in combination with eugenol (Guin *et al.*, 1984). The detailed mechanism of this quenching effect has not been elucidated, but it is believed to involve competitive inhibition at the receptor level.

A balanced approach is therefore required. Although the benefits of mono-substances are clearly recognized, the same attention should be paid to NCS to promote their research.

1.2.6   Re-evaluating NCS: A holistic approach to health care

Obviously, the tide is turning because the limited focus on mono-substances is leading to new problems. This may be for numerous reasons:

• **Economic viability:** Isolation and purification of single compounds is a costly endeavour. In addition, there is a growing demand for sustainable and environmentally friendly practices.
• **Patient-centred care:** There is an increasing demand for complementary and integrative healthcare approaches that consider the whole person

rather than only a single target or symptom. NCS, with their potential for multi-targeted action and reduced side effects, align well with this patient-centred approach (Agarwal, 2018; Clark *et al.*, 2021).

- **Ecological impact:** The environmental impact of large-scale extraction and synthesis of compounds is increasingly being questioned. The sustainable cultivation of medicinal plants to produce NCS has been identified as a more environmentally friendly approach.

By embracing a holistic perspective that considers both human and environmental health, the full potential of NCS can be unlocked and the way for a more sustainable and integrative approach to health care can be paved.

1.3   Essential Oils: Aromatic Extracts From Nature's Pharmacy

Essential oils occupy a prominent place among nature's therapeutic offerings and have a long and varied history of medicinal use (Plant *et al.*, 2019). These potent extracts, found in more than 17,500 aromatic plant species (Fig. 1.3; Regnault-Roger *et al.*, 2012), are produced and stored in various plant organs, i.e. blossoms (e.g. *Cananga odorata* (Lam.) Hook.F. & Thomson (Annonaceae), ylang-ylang), leaves (e.g. *Mentha longifolia* (L.) Huds. (Lamiaceae), horse mint), wood (e.g. *Santalum acuminatum* (R.Br.) A.DC. (Santalaceae), sandalwood), roots (e.g. *Peucedanum ostruthium* (L.) W.D.J.Koch (Apiaceae), masterwort), rhizomes (*Zingiber officinale* Roscoe (Zingiberaceae), ginger; *Curcuma longa* L. (Zingiberaceae), turmeric) and fruits (e.g. *Myristica fragrans* Houtt. (Myristicaceae), nutmeg; *Carum carvi* L. (Apiaceae), caraway). Defined as complex mixtures of secondary metabolites (Ahmad *et al.*, 2021), EOs derive their characteristic strong odours from a rich diversity of volatile compounds (Fig. 1.4), including terpenoids, aldehydes, alcohols, ketones and phenolics (Bakkali *et al.*, 2008; Sadgrove *et al.*, 2022). The European Chemicals Agency further clarifies that EOs are 'a volatile part of a natural product obtained by distillation, steam-distillation, or, in the case of citrus fruits, by squeezing', primarily composed of volatile hydrocarbons (ECHA, 2022; EDQM, 2024). This inherent complexity, with hundreds of individual components contributing to the overall aroma and therapeutic profile, positions EOs as prime examples of multi-component mixtures.

1.3.1   The multifaceted roles of essential oils in nature

In addition to their therapeutic benefits for humans, EOs play an important role in the plant kingdom (Fig. 1.5). In nature, they are the communication, interaction and protection tools between plants, as well as within their

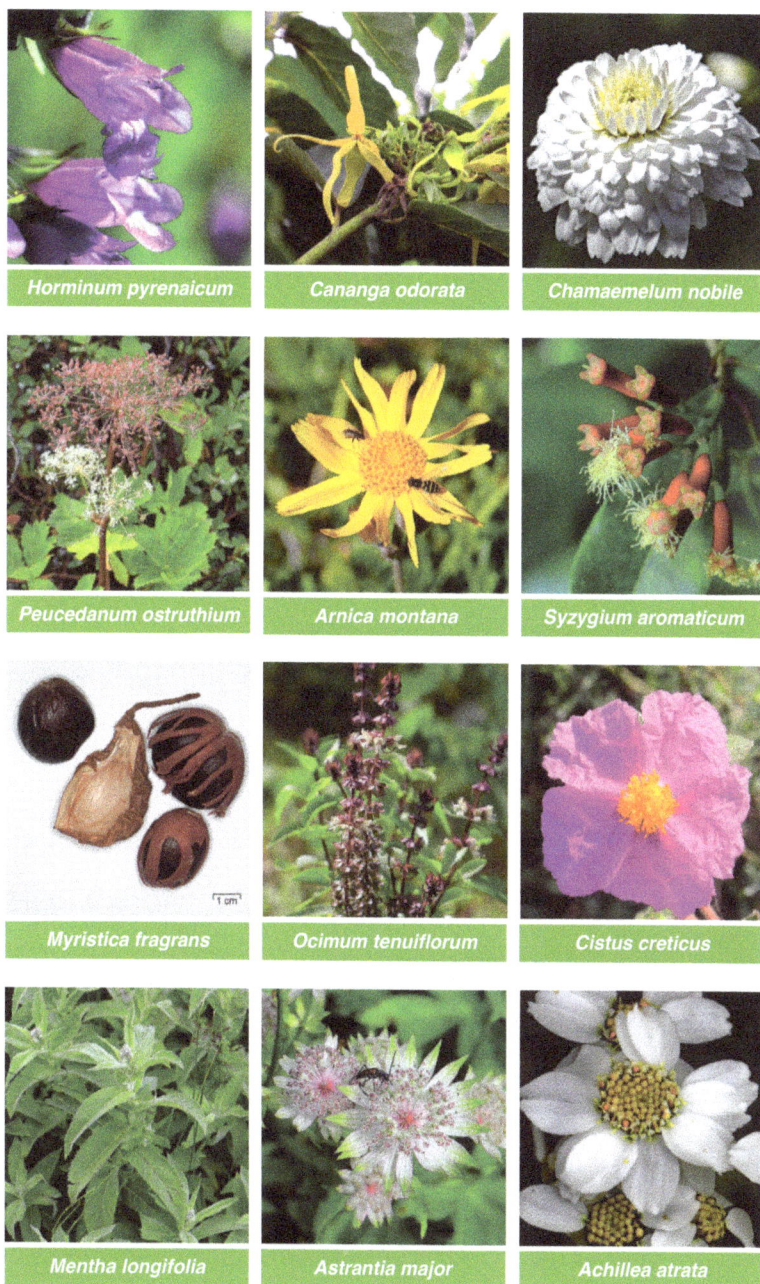

Fig. 1.3. Overview of some exotic and rare essential oil-containing plant species and plant parts. (Courtesy of Michael Keusgen.)

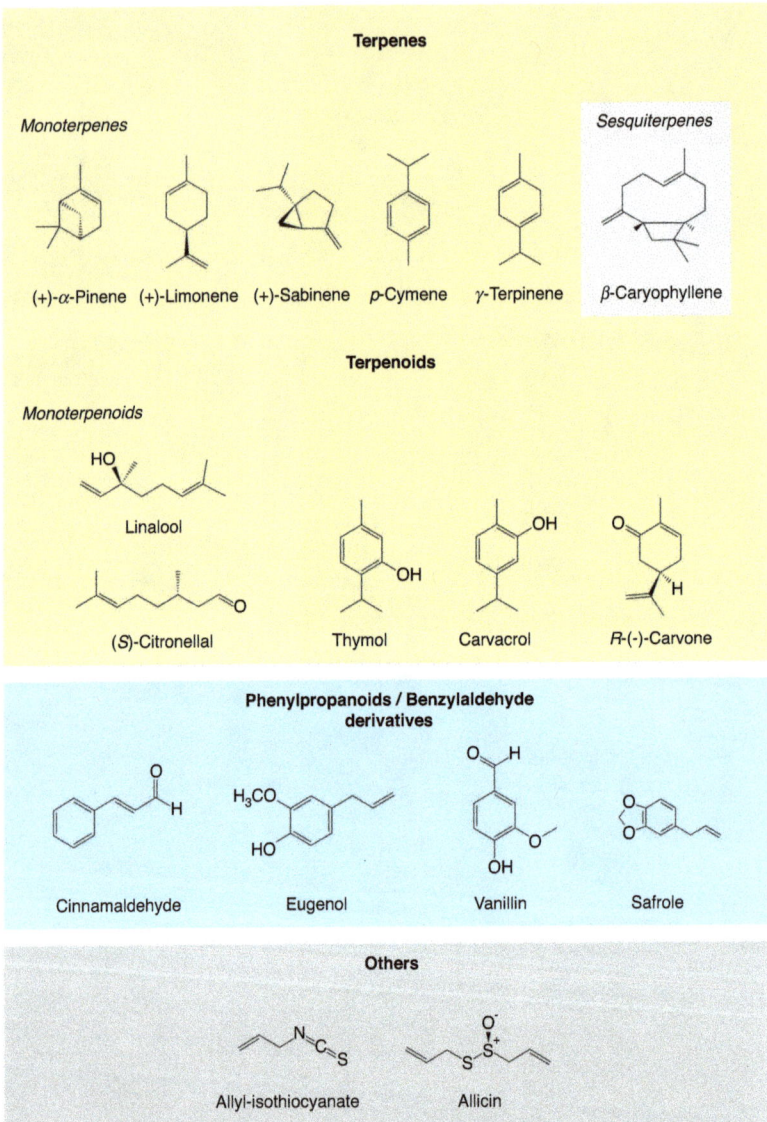

Fig. 1.4. Examples of chemical structures of essential oil constituents. (Adapted from Bunse *et al.*, 2022. CC BY.)

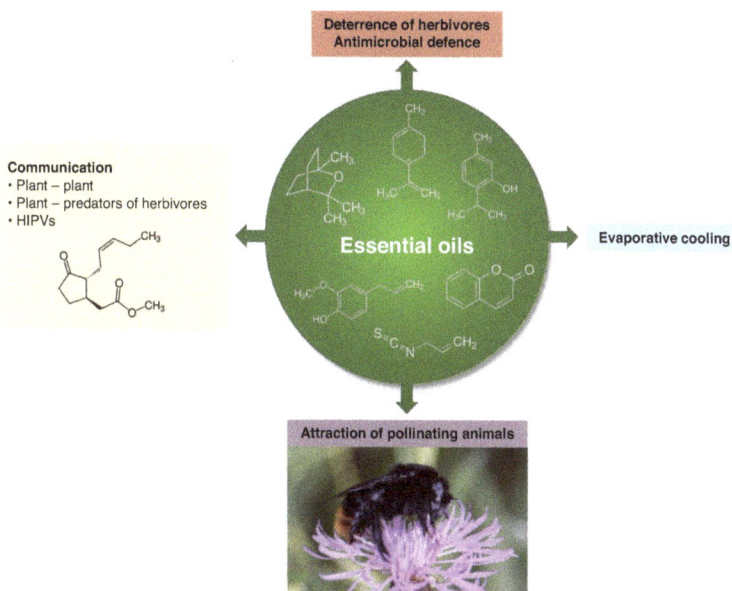

Fig. 1.5. Functions of plant fragrances – from the perspective of the plants. (Courtesy of Michael Wink.)

environment. Some important properties are defined as follows (Holopainen, 2004; Dudareva *et al.*, 2006; Loreto *et al.*, 2009; Loreto and D'Auria, 2022):

- **protectors:** shielding against pathogens like fungi and bacteria;
- **guardians:** deterring herbivores by attracting their natural enemies;
- **communicators:** sending signals to other plants, facilitating interspecies interactions;
- **attractors:** luring pollinators and seed dispersers, ensuring reproductive success; and
- **regulators:** helping plants adapt to environmental stressors like temperature fluctuations.

These complex interplays highlight the ecological significance of EOs and their various functions in plant ecosystems.

The composition of EOs is anything but static, and there are several factors that can cause significant variations in their quality, quantity and chemical profile. Even within the same plant species, the profiles of EOs can differ for the following reasons (Masotti *et al.*, 2003; Angioni *et al.*, 2006):

- **Extraction method:** Extraction techniques affect the compound profile and the quantities of extracted compounds.
- **Plant part:** Different plant organs yield EOs with distinct compositions.
- **Environmental factors:** Climate, soil conditions and even pest plague may impact EO profiles.

- **Plant age and stage:** The stage of development of the plants affects their EO production and composition.
- **Genetic variation:** Chemotypes, or chemical varieties within a species, can exhibit distinct EO profiles.

Furthermore, special cultivation methods and selective breeding of aromatic plants can produce different chemical profiles of EOs and new improved natural compositions (Toxopeus and Bouwmeester, 1992; Sarrou *et al.*, 2017; Lal *et al.*, 2018).

1.3.2 The aromatic alchemy: Biosynthesis of essential oil components

The captivating scents of aromatic plants arise from a complex symphony of volatile organic compounds of different origins (Fig. 1.6; Sangwan *et al.*, 2001; Caissard *et al.*, 2004):

1. **Shikimate and acetate malonate pathways:** These pathways give rise to phenolic compounds, contributing to the unique aromas and potential therapeutic properties of many EOs.
2. **Fatty acid derivatives:** These compounds, derived from fatty acids, add another layer of complexity to the chemical tapestry of EOs.
3. **Isoprenoids:** Forming the largest and most significant class of EO constituents, isoprenoids are responsible for the characteristic scents of many plants.

Terpenoids, the primary components of many EOs, are biosynthesized from isoprene units. These isoprene units are formed via two main pathways (Newman and Chappell, 1999):

- **Mevalonic acid pathway:** Occurring in the cytosol, this pathway is particularly active in the production of sesquiterpenes.
- **2-C-Methylerythritol-4-phosphate (MEP) pathway:** Taking place in plastids, this pathway is primarily responsible for monoterpene biosynthesis.

Isopentenyl diphosphate (IPP) and its isomer, dimethylallyl diphosphate (DMAPP), are the universal precursors for all terpenoids. These isomers are assembled in a head-to-tail manner:

- **monoterpenoids:** formed by the fusion of one IPP unit with its isomer DMAPP, resulting in geranyl diphosphate (GPP; Dubey *et al.*, 2003)); and
- **sesquiterpenoids:** created, when GPP is further elongated by another IPP unit, forming farnesyl diphosphate, a 15-carbon precursor (Blerot *et al.*, 2018).

While terpenes often take centre stage, EOs encompass a broader spectrum of chemical classes, including:

- **Aromatic compounds:** Characterized by their ring-shaped structures, these compounds significantly contribute to the distinctive aroma of many EOs.

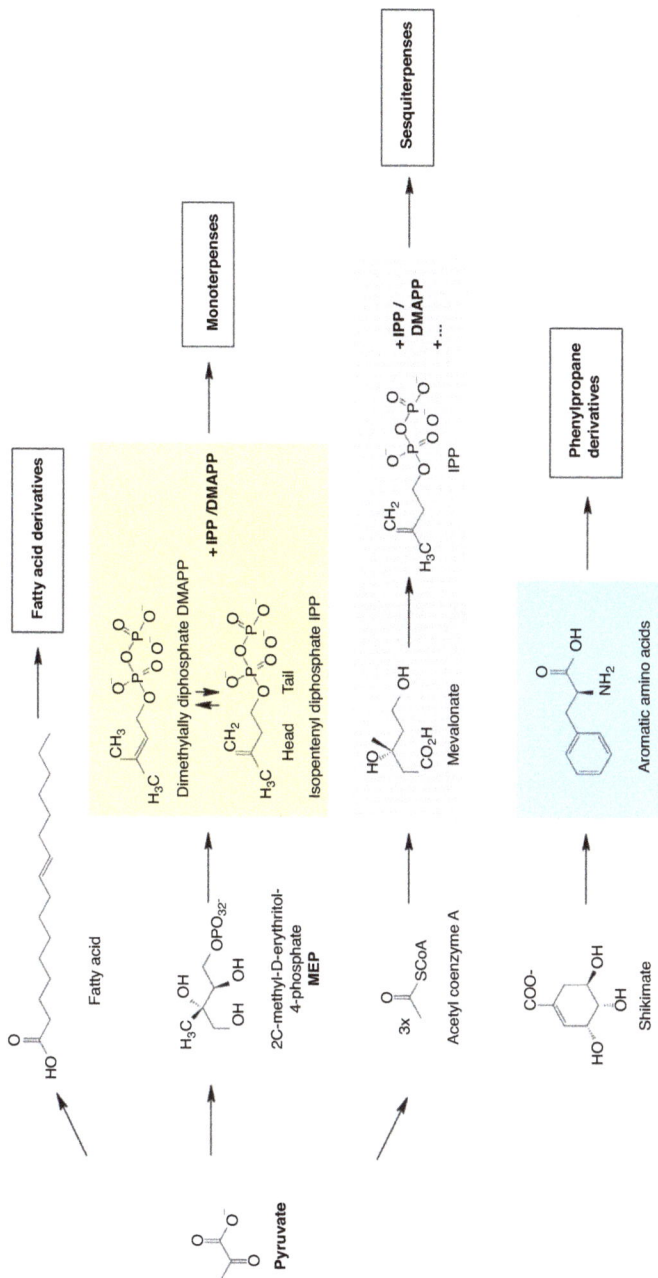

Fig. 1.6. Biosynthetic pathways of major volatile organic compounds (MEP: 2-C-methyl-erythritol-4-phosphate; IPP: isopentyl pyrophosphate). (Adapted from Bunse *et al.*, 2022. CC BY.)

- **Aliphatic compounds:** Composed of unbranched or branched carbon backbones, these constituents add further complexity to the chemical profile of EOs.

This diverse chemical palette, arising from distinct biosynthetic pathways, underscores the remarkable complexity and, among others, the therapeutic potential of EOs.

1.3.3   Unveiling the secrets of essential oils: Analytical techniques

Ensuring the quality, authenticity and safety of commercially available EOs requires sophisticated analytical techniques. Some of the most employed methods include:

- gas chromatography-mass spectrometry;
- headspace gas chromatography; and
- nuclear magnetic resonance spectroscopy.

These analytical methods play a crucial role in characterizing and standardizing EOs, ensuring their quality and safety for various applications. Further details on analytical and extraction techniques are explained in Chapter 2.

1.3.4   Essential oils: A global commodity with diverse applications

With approximately 3000 EOs from various aromatic plant species, these extracts have become a very important global commodity for numerous sectors (Fischman *et al.*, 2004; Ramezani *et al.*, 2008; Bento *et al.*, 2013; Turek and Stintzing, 2013; Seyyedi *et al.*, 2014; Ibrahim, 2020; Frix, 2023):

- **cosmetics and perfumes:** incorporating EOs for their fragrance and potential skin benefits. For further information, see Chapter 6;
- **food and flavourings:** enhancing the taste and aroma of processed foods. For further information, see Chapter 7;
- **sanitary and cleaning products:** leveraging the antimicrobial properties of EOs;
- **medicine and pharmaceuticals:** harnessing the therapeutic properties of EOs for various health conditions. For further information, see Chapter 4 (on veterinary medicine) and Chapter 5 (on clinical aromatherapy);
- **dentistry:** employing EOs for their antiseptic and analgesic properties; and
- **agriculture:** utilizing EOs as natural pesticides and plant protectants.

This widespread application highlights their economic and cultural significance and consolidates their position as valuable resources from nature.

1.3.5   Decoding the aromatic symphony: Essential oil composition

The unique fragrance of an EO is a complex interplay of multiple molecule structures; each compound may perform a different function that contributes to its overall aroma and therapeutic function as a natural complex substance (Stahl-Biskup and Reher, 1987). These complex blends usually contain the following:

- **Major constituents (20–95%):** These dominant compounds, often with low odour thresholds, largely define the characteristic scent of the EO.
- **Minor compounds (1–20%):** Present in smaller quantities, these compounds add subtle nuances and complexity to the aroma profile.
- **Trace compounds (<1%):** While present in minute amounts, these trace constituents can still influence the overall aroma and therapeutic activity.

This variability is illustrated by the differences in the chemical composition of various oils mentioned below (Asadollahi-Baboli and Aghakhani, 2015; Beigi *et al.*, 2018; Bhavaniramya *et al.*, 2019; Dobreva and Dimov, 2021; Poudel *et al.*, 2021):

- **oregano oil** (*Origanum vulgare* 'Compactum' L.; Lamiaceae): characterized by high levels of carvacrol (30–80%) and thymol (27–80%), whereas in fresh oregano leaves the main fragrance components are γ-terpinene, *p*-cymene, thymol and carvacrol;
- **coriander oil** (*Coriandrum sativum* L.; Apiaceae): dominated by linalool (68%), contributing to its sweet, floral aroma;
- **white wormwood oil** (*Artemisia herba-alba* Asso; Asteraceae): notable for its high content of α- and β-thujone (57%) and camphor (24%), historically used for its medicinal properties;
- **camphor tree oil** (*Cinnamomum camphora* (L.) J. Presl; Lauraceae): as its name suggests, this oil is rich in D-camphor (50%), known for its strong, penetrating aroma;
- **dill oil** (*Anethum graveolens* L.; Apiaceae): exhibiting variations in composition depending on the plant part, with the leaf oil containing high levels of α-phellandrene (up to 32%) and limonene (up to 32%), while the fruit oil is dominated by carvone (up to 55%) and limonene (up to 45%); and
- **peppermint oil** (*Mentha* × *piperita* L.; Lamiaceae): characterized by its refreshing aroma, attributed to significant amounts of menthol (up to 45%) and menthone (up to 15%).

1.3.6   The power of subtlety: Minor compounds, major impact

While major constituents often steal the spotlight, it is crucial to recognize the significant impact of minor and trace compounds in shaping the overall aroma profile of EOs. These seemingly insignificant components may possess remarkably low odour thresholds, i.e. they can be perceived even in minute quantities, adding depth and complexity to the scent (Góra and Brud, 1983), and are addressed as character impact substances.

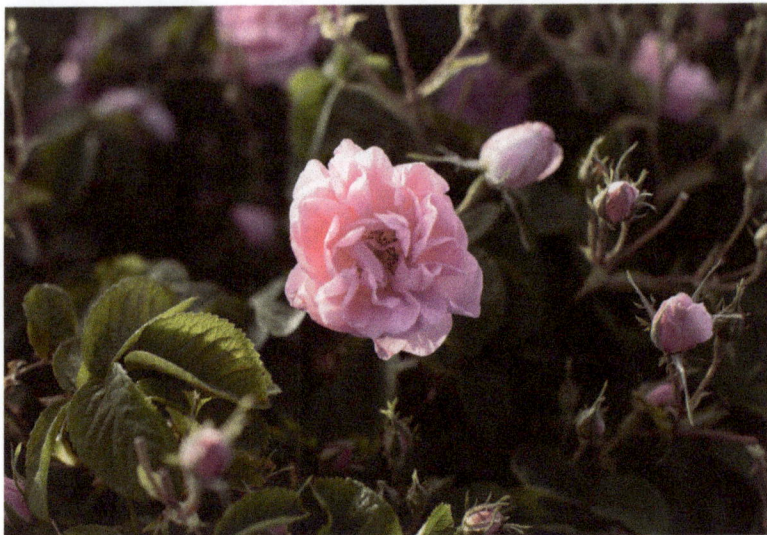

Fig. 1.7. Damask rose (*Rosa × damascena*). (Author's own image, courtesy of WALA Heilmittel GmbH.)

Case in point: Rose essential oil

Damask rose (*Rosa × damascena*; Fig. 1.7), with its captivating fragrance, is a fascinating example. About 27 compounds are characterized, whereas a handful of trace compounds, present at concentrations below 1%, are responsible for its distinctive, alluring scent. These potent trace compounds, including *β*-damascenone, rose oxide, *trans*-nerolidol, rotundone and 4-(4-methylpent-3 -en-1-yl)-2(5H)-furanone, demonstrate that even the smallest components can have a powerful impact on the olfactory experience (Naquvi *et al.*, 2014; Ohashi *et al.*, 2019).

These results underline a crucial aspect of EO analysis: A focus on the quantitative occurrence of compounds is not enough because all the complex interactions and aromatic depth cannot be illustrated. It is more important to analyse EOs in a more qualitative way, considering their odour thresholds and synergistic activities of all constituents, to unlock the full potential of EOs as natural aromatic mixtures.

1.4 Essential Oils and Their Bio-Functional Properties: A Double-edged Sword

Essential oils, known for their therapeutic potential, have a complex network of bio-functional properties derived from their diverse chemical composition. These properties with their beneficial effects can also cause risks if misused.

The remarkable molecular structures of each constituent are the reason for the various cellular targets and mechanisms of action. In contrast to designed remedies which are usually developed for a single mode of action, EOs exhibit a broader spectrum of activities. Disruption of the cellular membrane seems to be one main mechanism. EOs can penetrate cell walls and disrupt the structural integrity of membranes of bacteria, fungi and even mammalian cells. This disruption can alter the fluidity and permeability of membranes, influencing crucial cellular processes (Yoon *et al.*, 2000; Carson *et al.*, 2002; Armstrong, 2006).

1.4.1   Mitochondrial disruption and cytotoxicity

Mitochondria, the power stations of the cells, are particularly sensitive to EO activities, and disruption of their membranes can lead to:

- **decreased membrane potential:** impairing energy production;
- **disrupted ionic balance:** affecting calcium signalling and other vital processes; and
- **reduced pH gradient:** interfering with energy production and enzyme activity.

Such disruption can ultimately lead to cytotoxic effects that result in cell death through apoptosis, also known as programmed cell death, or necrosis, the uncontrolled cell death (Richter and Schlegel, 1993; Novgorodov and Gudz, 1996; Vercesi *et al.*, 1997).

1.4.2   Oxidative stress and bioenergetic failure

EOs can cause oxidative stress, a state of imbalance between free radicals and antioxidant strategies at the cellular level, which can lead to serious cellular damage, including DNA, proteins and lipids. Moreover, EOs may cause bioenergetic failure, reflecting a state of drained cellular energy, which aggravates the loss of viability (Vercesi *et al.*, 1997). This potential for disrupting cellular and mitochondrial integrity underlines the need for appropriate dilutions, methods of application and awareness of possible contraindications to minimize risks related to improper use.

In addition, the safety of EOs must be evaluated beyond possible cytotoxic effects and must include mutagenicity (inducing DNA mutation) and genotoxic (damaging DNA) activities.

1.4.3   Antimutagenic potential of essential oils

Research suggests that some EOs exhibit antimutagenic properties through various mechanisms (Ramel *et al.*, 1986; Kada and Shimoi, 1987; De Flora and Ramel, 1988; Hartman and Shankel, 1990; Kuo *et al.*, 1992; Shankel *et al.*, 1993; Waters *et al.*, 1996; Sharma *et al.*, 2001; Gomes-Carneiro *et al.*, 2005; Ipek *et al.*, 2005):

- **preventing mutagen entry:** acting as a barrier to protect cellular DNA;
- **direct scavenging:** neutralizing mutagens through chemical reactions;
- **antioxidant activity:** neutralizing reactive oxygen species that may damage DNA;
- **cellular antioxidant pathway activation:** boosting natural defence mechanisms of the cells against oxidative stress; and
- **metabolic modulation:** inhibiting the conversion of promutagens into active mutagens or enhancing the detoxification of harmful substances.

These multifaceted actions highlight the complex interplay between EOs and cellular processes involved in DNA protection.

1.4.4   Essential oils and DNA mutation: A nuanced perspective

Although EOs are generally considered safe, their potential effects on DNA integrity need to be carefully evaluated. Research in this area shows a complex situation with both reassuring and cautionary findings.

Numerous studies have found no evidence of nuclear DNA mutations induced by various EOs and constituents isolated therefrom (Bakkali *et al.*, 2005). However, for certain EOs and specific components there are indications of possible genotoxic effects in different *in vitro* test systems:

- **EOs with genotoxic potential:** Some *in vitro* studies have reported genotoxic activity of EOs from plants like *Artemisia dracunculus*, *Artemisia graveolens*, *Mentha spicata* and *Pinus sylvestris* (Zani *et al.*, 1991; Franzios *et al.*, 1997; Karpouhtsis *et al.*, 1998; Lazutka *et al.*, 2001).
- **Specific components of concern:** Compounds like *trans*-anethole, β-asarone, terpineol, *trans*-cinnamaldehyde, carvacrol, thymol and (*S*)-(+)-carvone have shown mutagenic potential in isolated form in the AMES test (Nestmann and Lee, 1983; Hasheminejad and Caldwell, 1994; Gomes-Carneiro *et al.*, 1998; Stammati *et al.*, 1999).

It must be noted that these *in vitro* studies often deal with higher concentrations than would be in normal administration. Furthermore, often only a single isolated compound was tested, and different mutagenic and genotoxic test systems may produce different results. Whether these genotoxic effects would occur with proper application remains questionable. The metabolic pathways of EOs make it even more complex. For example, some phenylpropanoids, which are very common in EOs, can be metabolized in the liver to potentially mutagenic epoxides (Wink and Schimmer, 2010). This is the reason why special plant varieties are cultivated, such as calamus (*A. calamus*) with low β-asarone contents (Bertea *et al.*, 2005).

Assessing the mutagenic and genotoxic potential of EOs requires a comprehensive approach considering:

- **Specific EO composition:** The risk level depends on the profile of individual constituents.
- **Dosage and application methods:** Intended use and dilution are crucial for minimizing potential harm.
- **Metabolic factors:** Understanding how EO components are metabolized *in vivo* is essential.
- **Long-term effects:** More research is needed to determine the long-term consequences of exposure to EO.

The therapeutic benefits of EOs can be utilized while ensuring their safe and responsible use, when these complexities are acknowledged and further research is promoted. Some EOs may have cytotoxic properties, while they often appear to be non-mutagenic, suggesting a generally low carcinogenic risk. It is important to differentiate that cytotoxicity does not mean carcinogenicity. However, some EO components might be 'secondary carcinogenic', i.e. they become carcinogenic when they are metabolized *in vivo*, e.g. in the liver. EOs like those from *Salvia sclarea* L. (Lamiaceae) and *Melaleuca quinquenervia* (Cav.) S.T.Blake (Myrtaceae) may influence hormone levels, potentially mimicking oestrogen. This has raised concerns about potential links to oestrogen-dependent cancers, although more research is needed to establish unequivocal connections (Guba, 2001). Moreover, oestrogen-active and endocrine-disrupting compounds need to be carefully differentiated (Stiefel and Stintzing, 2023). Also, when considering the impact of EOs on the human hormone system, the age-dependent vulnerability of individuals should be taken into account, as exemplified for lavender oil (Stiefel and Stintzing, 2023).

1.4.5 Photosensitization and phototoxicity

A key consideration is the occurrence of photosensitizing molecules in some EOs, such as those found in citrus oils. These molecules may – in the presence of sunlight (specifically UV-A) – cause skin damage, erythema (redness) and potentially increase skin cancer risk (Averbeck *et al.*, 1990; Averbeck and Averbeck, 1998; Nguyen *et al.*, 2020). Notably:

- **Furocoumarins:** These photosensitizing compounds are known to induce phototoxic effects and contribute to conditions like phytophotodermatitis.
- **Context-dependent effects:** It is important to note that these phototoxic effects are light-dependent. The same EO might not pose the same risk in the absence of UV exposure.
- **Non-phototoxic furocoumarins:** The majority of around 150 known furanocoumarins have not been sufficiently investigated. However, it has been shown that there are also non-phototoxic furanocoumarins such as bergaptol, isobergaptin and isopimpinellin (Scott *et al.*, 1976; Bitterling *et al.*, 2022a; Phucharoenrak and Trachootham, 2024).

Many *in vitro* studies (cell cultures) have been conducted to investigate the carcinogenicity of EOs, which may not accurately represent real-world conditions.

The concentrations used in these studies often exceed typical exposure levels. Thus, future research should focus on:

- ***in vivo* studies:** investigating long-term effects in living organisms to better understand real-world risks;
- **interrelationship of individual compounds:** examining how combinations of EO components and interactions with other substances might influence carcinogenicity; and
- **metabolic pathways:** elucidating the metabolic fate of EO constituents to identify potential carcinogenic metabolites.

Recent studies have pointed out the protective role of furanocoumarins against the oxidation of terpenes as shown by Bitterling and co-workers (Bitterling *et al.*, 2022a; Bitterling *et al.*, 2022b). This suggests that interactions among individual compounds of an EO, even between different types of natural complex substances such as EOs and polyphenols, might be much more important than understood hitherto and warrant further studies. While most EOs are not expected to be carcinogenic under typical conditions of use, a nuanced approach is essential. The understanding of specific EO compositions and the potential for metabolic activation, hormonal effects and phototoxic risks is very important for informed decision-making and safe application of EOs.

1.5   Future Directions

1.5.1   Scientific inquiry: Deepening our understanding

While traditional knowledge and anecdotal evidence have laid a strong foundation for EO application, further scientific inquiry is needed to validate efficacy and safety and to find new applications. Moreover, the knowledge generated through in-depth research can be used to better educate therapists and thus better match patient needs.

- **Unravelling complex synergies:** EOs represent complex mixtures of different constituents. Observation of the chemical interactions underlying synergy remains in its infancy, and as this relationship becomes more apparent in greater detail, specific clinical outcomes and successes will become better recognized.
- **Personalized aromatherapy:** The advances in understanding individual biochemistry and genetics will also lead to personalized aromatherapy. Imagine a future where EO blends are tailored to an individual's specific needs and responses, maximizing their therapeutic benefits.
- **Essential oils and the microbiome:** The mammalian gut microbiota plays an inherent role in health behaviour. When bacterial flora is out of balance, natural complex substances such as EOs may help bring it back into balance. Further investigation into this relationship may provide new uses for digestive health, immunity and mood management.

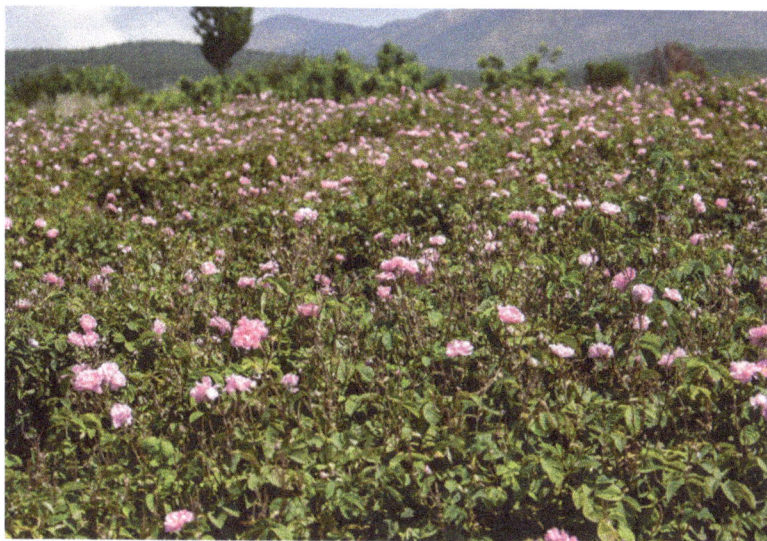

Fig. 1.8. Cultivation of Damask roses (*Rosa × damascena*) in Turkey. (Photo: Courtesy of Catrin Cohnen-Deliga.)

1.5.2 Expanding horizons: Essential oils in a global context

Sustainability and environmental responsibility have become more important in recent years. The increasing interest in natural wellness practices, natural cosmetics and integrative medicine has created new opportunities for ethical and sustainable sourcing of natural products such as EOs to develop a healthy world in the context of One Health.

- **Empowering local communities:** Supporting fair-trade practices and directly working with small-scale farmers and distillers empowers local communities while preserving traditional knowledge.
- **Sustainable harvesting and production:** The increasing demand for EOs occurs over time with increased demand for sustainable harvesting and production techniques. The protection of plant species and their ecosystems is as important as environmentally friendly harvesting and production methods (Fig. 1.8).
- **Bridging traditional and modern medicine:** Combining traditional evidence with up-to-date knowledge and more personal responsibility may have a great impact on new healthcare applications and a healthy mind. It would therefore not be surprising if the long evolving history of phytotherapy and integrative medicine forms an important bridge between these two camps – modern health.

Pharmacopoeias and Beyond – Analytical Methods for Assessment of Essential Oils, Proof of Specifications and Detection of Adulterations

2

Jörg Heilmann* and Gertrud E. Morlock

2.1 General Aspects of Analytical Assessment and Proof of Specifications

Essential oils (EOs) are complex mixtures with a large and naturally varying number of substances that are marketed in different formulations at a broad variety (Bunse *et al.*, 2022). The assessment of their qualitative and quantitative composition as well as their specifications regarding plant, plant part origin and processing requires powerful analytical methods to test and assess their content, purity, identity, stability and phytogenic nature and thus to adequately control and ensure the quality of EOs. The continuous monitoring of the regulatory and technical literature as well as the establishment and routine implementation of meaningful methods are indispensable for quality assurance and risk management. Several aspects of quality control are regulated in publicly accessible, validated official methods listed in the pharmacopoeias such as the German Pharmacopoeia (DAB 2023; Deutscher Apotheker Verlag, 2023), European Pharmacopoeia (PhEur 11th Edition; Bundesinstitut für Arzneimittel und Medizinprodukte, 2023), United States Pharmacopoeia combined with the National Formulary (USP-NF, 2023) and other monograph collections like the German Drug Codex (combined with *Neues Rezeptur Formularium*, DAC/NRF; ABDA, 2024). Various analytical techniques and methods therein allow both general testing of non-specific features and specific testing of qualitative, quantitative and biological characteristics. They include microbiological tests, organoleptic tests (such as taste,

*Corresponding author: joerg.heilmann@chemie.uni-regensburg.de

odour and bitter value), physico-chemical methods (such as relative density, refractive index and optical rotation), volumetric methods (e.g. hydrodistillation and drug-to-distillate ratio), spectrophotometric methods (UV/Vis/FLD) and chromatographic methods, e.g. thin-layer chromatography (TLC), high-performance thin-layer chromatography with multi-detection (HPTLC–UV/Vis/FLD) and gas chromatography coupled to flame ionization detection (GC–FID). Thus, the current analytical state of the art in pharmacopoeias focuses on physico-chemical data, volumetric content determination, quantification of value-giving constituents and specific marker compounds, limitation of probably toxic substances and chromatographic profiles of EOs. Interestingly, the versatile and powerful high-performance liquid chromatography (HPLC), which is easy to couple to various detectors (UV/Vis/FLD, photo diode array, mass selective detector, etc.) is not used for EOs in PhEur.

A further comprehensive analysis of such multi-compound mixtures is limited due to the otherwise considerable effort, the costs required and the lack of toxicity data (not available for all detected compound signals). Medicinally used EOs, which meet the quality criteria of PhEur are per definition natural (i.e. phytogenic). Synthetic EOs, despite being (largely) identical to phytogenic EOs (so called nature-identical EOs), are generally not permitted by PhEur. Nevertheless, various EOs are adulterated by the addition of synthetically produced 'nature-identical substances'. The analytical challenges of distinguishing genuine natural from nature-identical compounds/oils, and thus the proof of the entire phytogenic origin of an EO, require the use of more sophisticated analytical techniques and are based on enantioselectivity and isotope discrimination. Surprisingly, these methods are mostly beyond the scope of the above-mentioned pharmacopoeias, but an integral part of the quality control in industrial processes of food, flavours and fragrances.

2.2   Identification of Adulterants, Contaminants and Residues – General Aspects and Examples

Accurate and reliable methods are crucial for consumers to have confidence in the quality and authenticity of EOs. Not only value-determining and marker compounds, but also adulterants, residues or contaminants are in the analytical focus. Essential oils are comparatively expensive products and thus potentially susceptible to falsification, adulteration and contamination in increasingly global supply chains, often traded along various suppliers and sometimes even anonymous, with unclear upstream supply/manufacturing/production/process chains.

Non-intentional contamination can occur in various unknown ways, which is difficult to capture analytically. Falsification, which misrepresents the original quality or composition, can occur during plant collection, packaging, labelling (European Commission, 2011) and further processing steps. They may be recognized by clear actual deviations in appearance or smell from the reference.

Case in point: Adulteration during wild harvesting

A typical example of falsification of an EO plant during wild collection is the mistaken identity of different chamomile species (*Anthemis cotula* L., mayweed chamomile; *Tripleurospermum inodorum* (L.) SCH. BIP., scentless chamomile; *Chamaemelum nobile* (L.) ALL., roman chamomile; all Asteraceae) with *M. recutita* L. (Guzelmeric *et al.*, 2015), leading to products with reduced quality (*T. inodorum*) and/or changed composition (*A. cotula*, *Ch. nobile*). Whereas the EO of *M. recutita* contains highly anti-inflammatory active bisabolols, bisabolol oxides and chamazulene as main compounds, the EO of *A. cotula* contains mainly monoterpene and sesquiterpene hydrocarbons (germacrene-D and spathulenol; Saroglou *et al.*, 2006) and the EO of *Ch. nobile* consists of esters of short-chain alcohols (C_3 to C_6) with short-chain carboxylic acids (C_4/C_5), especially angelica acid (Wang *et al.*, 2014). Interestingly, both EOs are of insufficiently characterized pharmacological activity.

Adulteration is aimed at intentionally creating a non-genuine product that resembles the standard in terms of measurable value-giving characteristics by:

- dilution with cheaper oils;
- blending with lower quality materials;
- adding natural or synthetic compounds that are not naturally present in the EO; and
- utilizing poor-quality raw materials, waste material or by-products.

One form of adulteration is the dilution of essential oils with mineral oils (Bounaas *et al.*, 2018), synthetic or less expensive vegetable, fatty oils such as sunflower (Fig. 2.1) or soybean oil (Do *et al.*, 2015), which reduces the concentration of the EO, while the loss of scent can be offset by the addition of less costly synthetic fragrances. For example, adding synthetic linalool to lavender EO enhances its fragrance (Wang *et al.*, 2021) or adding palmarosa oil to damask rose oil increases its geraniol content (Do *et al.*, 2015). Recently samples of fennel, anise and star anise EO were analysed by GC-MS and detected to be adulterated by dilution with triethyl citrate and capryl palmitate (Murphy *et al.*, 2024). Another adulteration form involves blending EOs with lower quality and thus less expensive EOs from different botanical species or regions that possess similar chemical constituents to the target EO (Do *et al.*, 2015). Blending is also done to conceal an inferior quality of the original EO. The use of poor quality or contaminated raw materials, e.g. using non-distilled or chemically extracted oils or oils obtained from old or spoiled botanical material, can potentially be harmful. In any case, falsifications and adulterations alter the chemical profile and compromise the authentic and therapeutic properties that consumers seek in EOs. Nevertheless, since adulterations can be extremely diverse, very different analytical methods must also be used to detect them.

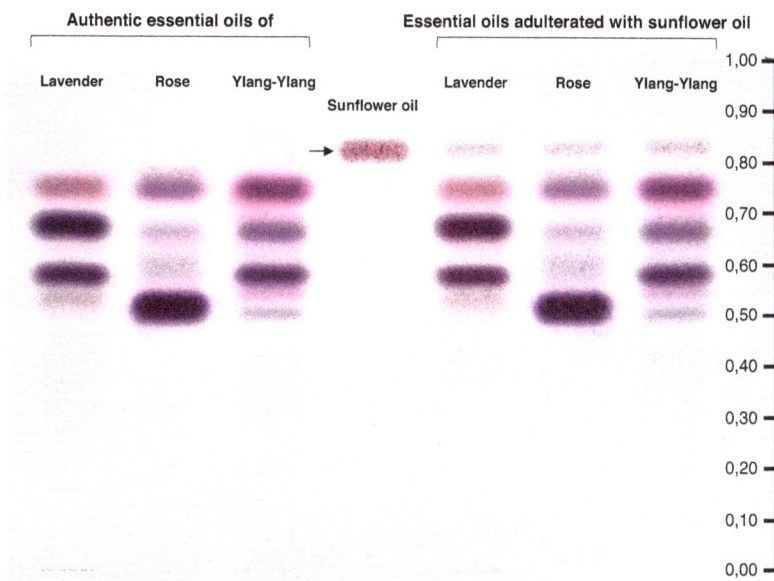

Fig. 2.1. Example of a TLC fingerprint of authentic (tracks 1–3) and with sunflower oil adulterated essential oils (tracks 5–7), derivatized with anisaldehyde sulfuric acid reagent and detected at Vis. (Author's own image.)

The American Botanical Council, through its Botanical Adulterants Program, has contributed to the systematic documentation of adulterations of EOs observed on the market and their analysis methods (American Botanical Council, n.d.). Apart from pharmacopoeia, requirements regarding sampling, methodology and validation of known adulterants (e.g. mineral oil-based flavour compounds, synthetic esters or aldehydes), contaminants (e.g. mycotoxins, polycyclic aromatic hydrocarbons or tropane alkaloids; European Commission, 2006) and residues such as pesticides (European Commission, 2005) or solvents (European Commission, 2009) are regulated on the European level where food fraud is prosecuted (European Commission, 2002, 2017). In addition, besides the EO monographs, there are validated methods on the federal level, exemplarily, for pyrrolizidine alkaloid analysis from the German Federal Institute for Risk Assessment (Bundesinstitut für Risikobewertung, BfR, 2014). Further examples of sources are the method collection according to §64 of the German Food and Feed Law (Lebensmittel- und Futtermittelgesetzbuch, 2021) as well as methods according to the Association of Official Agricultural Chemists (2023) and the International Organization for Standardization (2023). Otherwise, developed and validated in-house methods can be applied to an appropriate extent.

2.3   Standard Methods of the Pharmacopoeias – Targeting the Identity, Purity and Stability of EOs

In PhEur, the test for identity of EOs is uniformly regulated and based in all monographs on a TLC (test A) and a GC–FID-based method (test B). For TLC a silica gel (normal phase) plate is used as the stationary phase and the mobile phase is usually an ethyl acetate – toluene mixture in a variable volume ratio of 5:95, 10:90 or 15:85. In exceptional cases, ethyl acetate is substituted by methanol or the mobile phase changed to another lipophilic solvent. The more uniform character of the mobile phase is explained by the basically lipophilic properties of all EOs. For the stationary phase, the use of silica gel with a particle size diameter of 5–40 µm (TLC) as well as 2–10 µm (HPTLC) is permitted. The latter HPTLC allows the application quantities for test solutions and references to be reduced and the running distances to be shortened. This leads to reduced time and costs, although the stationary HPTLC phase is naturally more expensive. In many cases, HPTLC also improves separation performance (Reich and Schibli, 2011; Wilson and Poole, 2023). Due to the multi-substance nature of EOs, non-specific derivatization reagents like anisaldehyde or vanillin combined with sulfuric acid are preferred for detection. As references, often characteristic, value-determining or quantitatively dominating EO compounds of the respective oil but also distinguishing markers or retardation factor (R_F) markers are used. For profiling via such derivatization reagents, all experimental conditions must be strictly preserved. Due to the unspecific nature of the reaction (oxidation, dehydration and condensation), the detection colours are difficult to explain. The reproducibility of the coloration depends on numerous factors, in particular, type of the stationary phase, residuals of the mobile phase, drying time after the separation, derivatization time, derivatization temperature and amount of the compound (Jork *et al.*, 1989).

The identity test B in PhEur is the assignment of several EO substances in a GC–FID chromatogram, using mostly macrogol 20,000 as stationary phase. In deviation, phenyl(5)methyl(95)polysiloxane is used as stationary phase for analysis of Aurantii dulcis (*Citrus* x *sinensis* (L.) Osbeck, Rutaceae), Citri reticulatae (*C. reticulata* Blanco, Rutaceae), Thymi typo thymolo (*Thymus vulgaris* L., *T. zygis* L., Lamiaceae) and Juniperi (*Juniperus communis* L., Cupressaceae) aetheroleum. For implementation, a precisely defined reference solution from numerous characteristic substances present in the EO is mixed and chromatographed to characterize the chromatographic profile of the EO. The reference mixture and test solution are analysed and compared in different runs (external calibration). An example chromatogram is sometimes depicted in the respective monograph. Also, the usage of a Certified Reference Standard (CRS) EO can be described, as in the monograph of Spicae aetheroleum (from *Lavandula latifolia* Medik., Lamiaceae). Although it is used to determine identity, the GC–FID chromatogram was originally recorded as part of the purity test of EOs.

The preparation of precisely defined solutions of references allows as part of the purity section the quantification of the individual components via normalization. The PhEur distinguishes between limitation of an undesirable substance, minimum requirement, and defined content range for a substance. Quantification of the enantiomeric purity of single compounds is limited in PhEur to just a few EOs. For implementation, one must differentiate between an enantioselective chromatography (with chiral phases) or enantioselective detection or a combination of both. In PhEur, enantioselective chromatography using GC–FID with modified β-cyclodextrin as stationary phase is always applied. For Neroli aetheroleum (from *Citrus aurantium* L. ssp. *aurantium* L., Rutaceae) and Lavandulae aetheroleum (from *Lavandula angustifolia* MILL., Lamiaceae), the amount of (S)-(+)-linalool and (S)-(+)-linalyl acetate is limited. Similarly, the amount of (-)-carvone in Carvi aetheroleum (from *Carum carvi* L., Apiaceae) and the content of (R)-linalool in Coriandri aetheroleum (from *Coriandrum sativum* L., Apiaceae) are limited (Table 2.1).

Besides GC, also physico-chemical characteristics and methods of pharmacognosy are established for purity testing. The former includes relative density, refractive index and optical rotation. Relative density is given as d_{20}^{20} (weighed at 20°C), refractive index as n_D^{20} (D-line of sodium light $\lambda = 589.3$ nm) and optical rotation α in degree (°; at 20°C, D-line of sodium light $\lambda = 589.3$ nm). In exceptional cases, as for distinguishing Anisi aetheroleum (from *Pimpinella anisum* L., Apiaceae) from Anisi stellati aetheroleum (from *Illicium verum* HOOK.F., Illiciaceae), also the solidification temperature must be determined. The following section in PhEur contains, e.g. tests on the presence of water, foreign esters, fatty oils and resinified EO, solubility in ethanol or evaporation residue and the respective monograph describes which test needs to be carried out (Table 2.2).

Case in point: Resinous artefacts

As an example, the test for resinified EO is suitable for unstable EOs with polymerizing compounds. EOs containing monoterpenes with low oxygen functionalization tend to gum up under wrong storage conditions, with light, heat or oxygen forming resinous artefacts as in the EOs of Pinaceae and Cupressaceae like Pini sylvestris aetheroleum (from *Pinus silvestris* L.), Pini pumilionis aetheroleum (from *P. mugo* TURRA), Terebinthini aetheroleum (from *P. pinaster* AITON and/ or *P. massoniana* D.DON, all Pinaceae) and Juniperi aetheroleum. Thus, stability aspects are also considered in the purity tests.

Table 2.1. GC–FID of EOs in PhEur; chromatographic profile on Macrogol 20000 R as stationary phase (exception marked), quantitative value (determined by normalization) of main compounds, trace substances, critical compound pairs and toxicologically suspect compounds (permitted content in EO); GC–FID separation of chiral substances on modified β-cyclodextrin as stationary phase. (Authors' own table.)

Aetheroleum	Chromatographic profile	Compound with threshold	Critical compound pair (resolution)[a]	Toxicologically suspect compound
Anisi	linalool, estragole, α-terpineol, cis-anethole, trans-anethole (87–94%), anisaldehyde, pseudoisoeugenyl-(2-methylbutyrate)	fenchone ≤0.01%, foeniculin ≤0.01%, linalool ≤1.5%, α-terpineol ≤ 1.2%	estragole/α-terpineol (1.5)	estragole (0.5–5%)
Anisi stellati	linalool, estragole, α-terpineol, cis-anethole, trans-anethole (86–93%), anisaldehyde, foeniculin	fenchone ≤0.01%, pseudoisoeugenyl-2-methylbutyrate ≤0.01%, α-terpineol ≤ 0.3%	estragole/α-terpineol (1.5)	estragole (0.5–6.0%)
Aurantii dulcis	α-pinene, sabinene, β-pinene, β-myrcene, octanal, limonene (92–97%), linalool, decanal, neral, geranial, valencene		sabinene/β-pinene (1.5)a	
Carvi	β-myrcene, limonene, trans-dihydrocarvone ≤2.5%, carvone (50–65%), trans-carveol	trans-dihydrocarvone ≤2.5% trans-carveol ≤2.5%	β-myrcene/limonene (4.5)	
		(-)-/(+)-carvone, whereby (-)-carvone ≤1.0%	(-)-/(+)-carvone (2.4)	
Caryophylli floris	β-caryophyllene, eugenol (75–88%), acetyleugenol		eugenol/acetyleugenol (1.5)	

Continued

Cinnamomi cassiae	*trans*-cinnamyl aldehyde (70–90%), cinnamyl acetate, eugenol, *trans*-methoxy cinnamyl aldehyde	eugenol ≤0.5%	*trans*-methoxycinnamyl aldehyde/cumarine (1.5)
Cinnamomi zeylanici corticis	cineol, linalool, β-caryophyllene, safrol, *trans* cinnamyl aldehyde (55–75%), eugenol, cumarine, *trans* 2-methoxy cinnamyl aldehyde, benzyl benzoate	cineol ≤3.0% safrol ≤0.5% eugenol ≤7.5% cumarine ≤0.5% benzyl benzoate ≤1.0%	linalool/β-caryophyllene safrol (1.5)
Cinnamomi zeylanici folii	cineol, linalool, β-caryophyllene, safrol, *trans* cinnamyl aldehyde, cinnamyl acetate, eugenol (70–85%), cumarine	cineol ≤1.0% safrol ≤3.0% *trans*-cinnamyl aldehyde ≤3.0% cinnamyl acetate ≤2.0% cumarine ≤1.0%	linalool/β-caryophyllene safrol (1.5)
Citri reticulatae	α-pinene, sabinene, β-pinene, β-myrcene, p-cymene, limonene (65–75%), γ-terpinene, methyl-(N-methylanthranilat)	sabinene ≤0.3% p-cymene ≤1.0%	sabinene/β-pinene (1.5)[a] p-cymene/limonene (1.5)[a]
Citronellae	limonene, citronellal (30–45%), citronellyl acetate, neral, geranial, geranyl acetate, citronellol, geraniol	neral ≤2.0% geranial ≤2.0%	geranyl acetate/citronellol (1.2)
Coriandri	α-pinene, limonene, γ-terpinene, camphor, p-cymene, linalool (65–78%), α-terpineol, geranyl acetate, geraniol	(R)-/(S)-linalool, whereby (R)-linalool ≤14%	linalool/camphor (1.5) (R)-/(S)-linalool (5.5) (S)-linalool/borneol (2.9)

Table 2.1. Continued

Aetheroleum	Chromatographic profile	Compound with threshold	Critical compound pair (resolution)[a]	Toxicologically suspect compound
Eucalypti	α-pinene, β-pinene, sabinene, α-phellandrene, limonene, 1,8-cineol (≥ 70%), camphor	sabinene ≤0.3% camphor ≤0.1%	limonene/1,8-cineol (1.5)	
Foeniculi amari fructus	α-pinene, limonene, fenchone, estragole, cis-anethole, trans-anethole (55–75%), anisaldehyde	cis-anethole ≤0.5% anisaldehyde ≤2.0%	estragole/trans-anethole (5.0)	estragole ≤6.0%
Foeniculi amari herbae Spanish type	α-pinene, β-pinene, β-myrcene, α-phellandrene, limonene, fenchone, estragole, cis-anethole, trans-anethole (15–40%), anisaldehyde, anisketone	cis-anethole ≤0.5% anisaldehyde ≤1.0% anisketone ≤0.05%	β-myrcene/α-phellandrene (1.5)	estragole (2.0–7.0%)
Tasmanic type	α-pinene, α-phellandrene, limonene, fenchone, estragole, cis-anethole, trans-anethole (15–40%), anisaldehyde, anisketone			estragole (1.5–6.0%)
Juniperi	α-pinene (20–50%), sabinene, β-pinene, β-myrcene (1–35%), α-phellandrene, limonene, terpinen-4-ol, bornyl acetate, β-caryophyllene	sabinene ≤20% α-phellandrene ≤1.0% bornyl acetate ≤2.0% β-caryophyllene ≤7.0%	sabinene/β-pinene (1.5)[a]	

Continued

Lavandulae	limonene, 1,8-cineol, 3-octanone, camphor, linalool (20–45%), linalyl acetate (25–47%), terpinen-4-ol, lavandulol, α-terpineol	limonene ≤1.0% 1,8-cineol ≤2.5% camphor ≤1.2% α-terpineol ≤2.0%	terpinen-4-ol/lavandulyl acetate (1.4)
		(R)-/(S)-linalool, whereby (S)-linalool ≤12% (R)-/(S)-linalyl acetate, whereby (S)-linalyl acetate ≤1%	(R)-/(S)-linalool (5.5) (R)-/(S)-linalyl acetate (2.0) (S)-linalool/borneol (2.9)
Limonis	β-pinene, sabinene, limonene (56–78%), γ-terpinene, β-caryophyllene, neral, α-terpineol, neryl acetate, geranial, geranyl acetate	β-caryophyllene ≤0.5% α-terpineol ≤0.6%	β-pinene/sabinene (1.5), geranial/geranyl acetate (1.5)
Matricariae rich in bisabolol oxides rich in (-)-α-bisabolol	(-)-α-bisabolol, chamazulene, guajazulene bisabolol oxides (29–81%) (-)-α-bisabolol (10–65%), bisabolol oxides + (-)-α-bisabolol (≥ 20%) chamazulene (≥ 1% both)		chamazulene/ guaiazulene (1.5)
Melaleucae	α-pinene, sabinene, α-terpinene, limonene, cineol, γ-terpinene, p-cymene, terpinolene, terpinen-4-ol (≥ 30%), aromadendrene, α-terpineol	sabinene ≤3.5% cineol ≤15.0% aromadendrene ≤7.0%	terpinen-4-ol/ aromadendrene (2.7)
Menthae arvensis aetheroleum partim mentholum depletum	limonene, cineol, menthone (17–35%), isomenthone, mentyl acetate, isopulegol, menthol (30–50%), pulegone, carvone	cineol ≤1.5% pulegone ≤2.5% carvone ≤2.0% ratio cineol/ limonene < 1	cineol/limonene (1.5)

Table 2.1. Continued

Aetheroleum	Chromatographic profile	Compound with threshold	Critical compound pair (resolution)[a]	Toxicologically suspect compound
Menthae ×piperitae	limonene, 1,8-cineol, menthone (14–32%), menthofuran, isomenthone, menthyl acetate, isopulegol, menthol (30–55%), pulegone, carvone	isopulegol ≤0.2%, pulegone ≤3.0% carvone ≤1.0% ratio cineol/ limonene > 2	limonene/1,8-cineol (1.5) piperitone/carvone (1.5)	
Myristicae fragrantis	α-pinene (15–28%), β-pinene (13–18%), sabinene (14–29%), car-3-ene, limonene, γ-terpinene, terpinen-4-ol, safrol, myristicin	safrol ≤2.5%	β-pinene/sabinene (1.5)	safrol, myristicin (5.0–12.0%)
Neroli	β-pinene, limonene, linalool (28–44%), linalyl acetate, α-terpineol, neryl acetate, geranyl acetate, trans-nerolidol, methylanthranilate, (E,E)-farnesol	neryl acetate ≤2.5% (R)-/(S)-linalool, whereby (S)-linalool ≤30% (R)-/(S)-linalyl acetate, whereby (S)-linalyl acetate ≤5%	β-pinene/sabinene (1.5) (R)-/(S)-linalool (5.5) (R)-/(S)-linalyl acetate (2.7)	
Niaouli typo cineolo	α-pinene, β-pinene, limonene, 1,8-cineol (45–65%), p-cymene, benzaldehyde, α-terpineol, trans-nerolidol, viridiflorol	methyleugenol ≤0.05% isomethyleugenol ≤0.05%	limonene/1,8-cineol (1.5)	
Pini pumilionis	α-pinene (10–30%), camphene, β-pinene, car-3-ene (10–40%), β-myrcene, limonene, β-phellandrene, p-cymene, terpinolene, bornyl acetate, β-caryophyllene	camphene ≤2.0% p-cymene ≤2.5% terpinolene ≤8.0%	car-3-ene/β-myrcene (1.5)	
Pini silvestris	α-pinene (32–60%), camphene, β-pinene, car-3-ene, β-myrcene, limonene, β-phellandrene, p-cymene, terpinolene, bornyl acetate, β-caryophyllene	β-phellandrene ≤2.5% p-cymene ≤2.0% terpinolene ≤4.0%	car-3-ene/β-pinene (1.5)	

	Composition	Limits	Ratio	
	α-pinene (18–26%), camphene, β-pinene, β-myrcene, limonene, cineol (16–25%), p-cymene, camphor (13–26%), bornylacetate, α-terpineol, borneol, verbenone		limonene/cineol (1.5)	
Rosmarini Spanish type Tunisian/ Moroccan type	α-pinene, camphene, β-pinene, β-myrcene, limonene, cineol (38–45%), p-cymene, camphor, bornyl acetate, α-terpineol, borneol, verbenone	verbenone 0.7–2.5% verbenone ≤0.4%	α-terpineol/borneol (1.5)	
Salviae lavandulifoliae	α-pinene, sabinene, limonene, 1,8-cineol (10–30.5%), thujone, camphor (11–36%), linalool, linalyl acetate, terpinen-4-ol, sabinyl acetate, α-terpinyl acetate, borneol	thujone ≤0.5% linalyl acetate ≤5.0% terpinen-4-ol ≤2.0%	limonene/1,8-cineol (1.5) α-terpinyl acetate/borneol (1.5)	thujone
Salviae sclareae	α-/β-thujone, linalool, linalyl acetate (56–78%), α-terpineol, germacrene-D, sclareol	α-/β-thujone ≤0.2% α-terpineol ≤5.0%	linalool/linalyl acetate (1.5)	α-/β-thujone
Spicae	limonene, 1,8-cineol (16–39%), camphor, linalool (34–50%), linalyl acetate, α-terpineol, trans-α-bisabolene	linalyl acetate ≤1.6%	limonene/1,8-cineol (1.5)	
Terebinthinae	α-pinene (70–85%), camphene, β-pinene, car-3-ene, β-myrcene, limonene, longifolene, β-caryophyllene, caryophyllene oxide	car-3-ene ≤1% caryophyllene oxide ≤1%	car-3-ene/β-myrcene (1.5)	
Thymi typo thymolo	α-thujene, β-myrcene, α-terpinene, p-cymene, γ-terpinene, linalool, terpinen-4-ol, carvacrol methylester, thymol (37–55%), carvacrol		thymol/carvacrol (1.5)[a]	

[a]Macrogol 20000 R is normally used as stationary phase, otherwise material is given in the table.

Table 2.2. Monographs on essential oils in PhEur; physico-chemical data, identity on TLC and methods of pharmacognosy. (Authors' own table.)

Aetheroleum	Relative density	Refractive index	Optical rotation	Solidification temperature	Identity[a,b]	Methods of pharmacognosy
Anisi	0.980–0.990	1.552–1.561		15–19°C	TLC: EtOAc:Tol 5:95	fatty oils and resinified essential oils
Anisi stellati	0.979–0.985	1.553–1.556		15–19°C	TLC: EtOAc:Tol 7:93, methyl-4-acetylbenzoate	fatty oils and resinified essential oils
Aurantii dulcis	0.842–0.850	1.470–1.476	+9.4° – +9.9°		TLC: EtOAc:Tol 15:85; UV 365 nm Absence of bergaptene	fatty oils and resinified essential oils
Carvi	0.904–0.920	1.484–1.490	+65° – +81°		TLC: EtOAc:Tol 5:95	
Caryophylli floris	1.030–1.063	1.528–1.537	–2° – 0°		TLC: Tol	fatty oils and resinified essential oils; solubility in ethanol
Cinnamomi cassiae	1.052–1.070	1.600–1.614	–1° – +1°		TLC: MeOH:Tol 10:90	
Cinnamomi zeylanici corticis	1.000–1.030	1.572–1.591	–2° – +1°		TLC: MeOH:Tol 10:90	
Cinnamomi zeylanici folii	1.030–1.059	1.527–1.540	–2.5° – +2.0°		TLC: MeOH:Tol 10:90	

Citri reticulatae	0.848–0.855	1.474–1.478	+6.4° – +7.5°	TLC: EtOAc:Tol 15:85; molybdatophosphoric acid	fatty oils and resinified essential oils; evaporation residue
Citronellae	0.881–0.895	1.463–1.475	−4° – +1,5°	TLC: EtOAc:Tol 10:90	
Coriandri	0.860–0.880	1.462–1.470	+7° – +13°	TLC: EtOAc:Tol 5:95	
Eucalypti	0.906–0.927	1.458–1.470	0° – +10°	TLC: EtOAc:Tol 10:90	solubility in ethanol
Foeniculi amari fructus	0.961–0.975	1.528–1.539	+10.0° – +24.0°	TLC: EtOAc:Tol 5:95, molybdatophosphoric acid	
Foeniculi amari herbae *Spanish type*	0.877–0.921	1.487–1.501	+42° – +68°	TLC: EtOAc:Tol 5:95, molybdatophosphoric acid	solubility in ethanol
Tasmanic type	0.940–0.973	1.512–1.538	+11° – +35°		
Juniperi	0.857–0.876	1.471–1.483	−15° – −0.5°	TLC: EtOAc:Tol 5:95	fatty oils and resinified essential oils
Lavandulae	0.878–0.892	1.455–1.466	−12.5° – −6.0°	TLC: EtOAc:Tol 5:95	
Limonis	0.850–0.858	1.473–1.476	+5.7° – +7.0°	TLC: EtOAc:Tol 15:85; UV without spray reagent at 366 nm	fatty oils and resinified essential oils; evaporation residue
Matricariae *rich in bisabololoxides* rich in *(-)-α-bisabolol*				TLC: EtOAc:Tol 5:95	

Continued

Table 2.2. Continued

Aetheroleum	Relative density	Refractive index	Optical rotation	Solidification temperature	Identity[a,b]	Methods of pharmacognosy
Melaleucae	0.885–0.906	1.475–1.482	+5° – +15°		TLC: EtOAc:Tol 20:80	
Menthae arvensis aetheroleum partim mentholum depletum	0.888–0.910	1.456–1.470	−34.0° – −16.0°		TLC: EtOAc:Tol 5:95	fatty oils and resinified essential oils
Menthae ×piperitae	0.900–0.916	1.457–1.467	−30° – −10°		TLC: EtOAc:Tol 5:95	
Myristicae fragrantis	0.885–0.905	1.475–1.485	+8° – +18°		TLC: EtOAc:Tol 5:95, vanilline	
Neroli	0.863–0.880	1.464–1.474	+1.5° – +11.5		TLC: EtOAc:Tol 15:85, 365 nm, absence of bergaptene	
Niaouli typo cineolo	0.904–0.925	1.463–1.472	−4° – +1°		TLC: EtOAc:Tol 5:95	
Pini pumilionis	0.857–0.870	1.474–1.480	−15° – −6°		TLC: EtOAc:Tol 5:95	fatty oils and resinified essential oils
Pini silvestris	0.855–0.875	1.465–1.480	−30° – −7°		TLC: EtOAc:Tol 5:95	fatty oils and resinified essential oils

Rosmarini Spanish type Tunisian/Moroccan type	0.895–0.920	1.464–1.473	−5° – +8°	TLC EtOAc:Tol 5:95, vanilline;	solubility in ethanol
Salviae lavandulifoliae	0.907–0.932	1.465–1.473	+7° – +17°	TLC: EtOAc:Tol 5:95, molybdatophosphoric acid	
Salviae sclareae	0.890–0.908	1.456–1.466	−26° – −10°	TLC: EtOAc:Tol 5:95, vanilline;	
Spicae	0.894–0.907	1.461–1.468	−7° – +2°	TLC: EtOAc:Tol 5:95,	solubility in ethanol fatty oils and resinified essential oils; evaporation residue
Terebinthinae	0.856–0.872	1.465–1.475	−40° – −28°	TLC: EtOAc:Tol 5:95	
Thymi typo thymolo	0.915–0.935	1.490–1.505		TLC: DCM	

[a]Anisaldehyde reagent is normally used as spray reagent for detection of essential oils on TLC. Otherwise, the detection reagent is given in the table.

[b]Part B of identity is always chromatographic profiling via GC–FID (Table 2.1).

2.4   Special Analytical Features of Essential Oils in PhEur – Implementations, Intentions and Limitations

Essential oils vary quantitatively and qualitatively not only in cases of different stem plants (Moore *et al.*, 2025) or geographic origins (Allenspach *et al.*, 2020) but also within the same region (Romanenko *et al.*, 2022). Consequently, PhEur has generally defined only ranges for the values of both, the physicochemical data (Table 2.2) and the content of individual substances determined via GC–FID chromatographic profiles (Table 2.1). For some EOs, however, it has proven necessary to define different geographic or chemical types due to variation in their analytical key values. An example for different origin definitions of an EO in one monograph is Foeniculi amari herbae aetheroleum (from *Foeniculum vulgare* MILL. ssp. *vulgare* var. *vulgare*, Apiaceae), with a Spanish and a Tasmanic type. The two types differ in all physico-chemical parameters as well as in the chromatographic profile. For example, the PhEur allows 15–40% *trans*-anethole and 8–30% limonene for the Spanish type, whereas the Tasmanic type must contain 45–76% *trans*-anethole and 1–6% limonene. However, the physico-chemical parameters can be identical in other EOs and only the chromatographic profiles differ. This is the case in the monograph Rosmarini aetheroleum (from *Salvia rosmarinus* SPENN., Lamiaceae) with a Spanish type exhibiting a high verbenone content of 0.7–2.5% and a Moroccan/Tunesian type with low verbenone content (≤ 0.4%). One example for an EO with different chemical types is Matricariae aetheroleum (*M. chamomilla* L., Asteraceae). To fulfil the chemical profile criteria, Matricariae aetheroleum rich in bisabolol oxides must contain 29–81% bisabolol oxides, whereas Matricariae aetheroleum rich in (-)-α-bisabolol needs 10–65% (-)-α-bisabolol and a total content of ≥20% (-)-α-bisabolol and bisabolol oxides. In contrast, PhEur claims for both types the identical content of chamazulene (≥ 1%).

2.4.1   Implementation of chromatographic profiles by GC

Chromatographic profiling of extracts can be understood as an analytical attempt to examine and characterize a multi-component mixture as well as possible with a chromatographic method (Wolfender *et al.*, 2015). In the best case, the characterization would realize the unmistakable identification of all peaks in a chromatogram, meaning that they all would be assigned to a respective known compound. Nevertheless, this requirement can hardly be met for multi-component plant extracts and thus chromatographic characterization should also be possible by the presence or absence of characteristic peaks without explicit assignment to a compound, what is called chromatographic fingerprinting (Kharbach *et al.*, 2020). The application of a suitable chromatographic method coupled to an appropriate detection ideally provides for the EO a typical chromatographic profile, an unmistakable fingerprint or a combination of both.

Recording a GC–FID-based chromatogram, PhEur aims to generate a quantitative chromatographic profile of an EO containing information on the identity, purity and content of individual substances. The aim of a pharmacopoeia is to ensure a comprehensive quality control by making the selection of reference compounds as characteristic as possible, and due to the high prices of references, to require as few substances as possible. Thus, the compounds chosen for quantitative profiling can be divided into three groups:

- quality-defining and -reducing compounds;
- (analytical) marker compounds; and
- compounds with toxicological background.

Due to the complex definition of various types of marker compounds, it should be noticed that the term 'quality-defining compounds' is used here instead of 'active marker compounds' as the latter term implies an evaluated efficacy, which is often not done on the single compound area for EOs. The marker compounds will be here understood as compounds (i) characterizing a species or a product obtained from a species (distinguishing marker), (ii) characterizing an adulteration or (iii) validating a method or system analytically.

2.4.1.1 Quantification of quality-defining and quality-reducing compounds

In EOs, quality-giving and quantitatively dominating substances can be identical but are not necessarily so. For example, the quality of anise EO depends upon high values of quantitatively dominating *trans*-anethole, whereas anethole oxidized forms such as anisaldehyde, anisketone and anisic acid are quality reducing. Also, the quality of clove EO (*Syzygium aromaticum* (L.) MERR. et L.M. PERRY, Myrtaceae) mainly depends on the main compound eugenol and in parts also on the acetyl eugenol content.

In most EOs, the situation is more complex as the quality mostly relies on the proportionate balance of various components. As an example, in EO of *Mentha × piperita* especially, concentrations of menthol, menthone, menthofuran and pulegone play a decisive role in quality assessment as quality is improved with higher menthol and menthone as well as lower menthofuran and pulegone content (Hudz *et al.*, 2023; Wang *et al.*, 2024). In neroli oil, linalool is the main compound of EO, but scent and aroma are influenced by several other monoterpenes (hydrocarbons and oxidized forms) such as β-pinene, limonene, geranyl acetate, linalyl acetate and *trans*-nerolidol (Cuchet *et al.*, 2021).

Worth noting is the fact that the analytical quality criteria of EOs in a pharmacopoeia do not always reflect the demands for other applications such as cosmetics or body care. In lavender oil (*Lavandula angustifolia* MILL., Lamiceae) different camphor-to-linalool ratios are associated with different usages (Wainer *et al.*, 2022), whereas PhEur limited camphor to 1.2%, and called for high linalool (20–45%) and linalyl acetate (25–47%) amounts for high-quality oils in therapy.

2.4.1.2   Marker compounds

Distinguishing markers are characteristic compounds for one species and are necessary to discriminate two EOs of very similar composition. Typical examples are the EOs of anise and star anise, which have the same main component *trans*-anethole in similar quantities (~90%), in addition to a largely congruent substance profile consisting of linalool, estragole, α-terpineol and *cis*-anethole. They can be distinguished by the substance foeniculin, which only occurs in star anise, and pseudoisoeugenyl-(2-methylbutyrate), which only occurs in anise. Accordingly, these substances are included in the respective limit values (foeniculin ≤0.01% in anise and pseudoisoeugenyl-(2-methylbutyrate) ≤ 0.01% in star anise) and their presence indicates a corresponding adulteration. Fenchone is also borderline restricted in both anise and star anise EO as it is an authentic marker compound for fennel EO and therefore suitable for recognizing an adulteration with fennel EO.

Since the analytical suitability test is part of every monograph, the chromatographic resolution of the GC system is validated by separation of a critical substance pair (Table 2.1). Chromatographic resolution is a parameter of how well two components are separated from each other in a chromatographic system. For a quantitative determination, a resolution of 1.5–2 is usually called for. However, this range is not always achieved in PhEur. For example, the critical pair of geranyl acetate and citronellol in Citronella aetheroleum (*Cymbopogon winterianus* Jowitt, Poaceae) provides a resolution of 1.2. The very similar retention times of both compounds in different GC systems (Anwar and Siringoringo, 2020) indicate that this is more a selectivity problem (retention difference) than an efficiency problem (peak broadening). For the critical pair of terpinene-4-ol and lavandulyl acetate in Lavandula aetheroleum, a resolution of 1.4 is required. It should be noticed that the question 'What is a critical pair?' strongly depends on the stationary phase used. For example, Pokajewicz *et al.* (2022) published a much better separation of both compounds in EO of different lavender cultivars using a 5% diphenyl- and 95% dimethyl-polysiloxane stationary phase; nevertheless, here lavandulol/terpinen-4-ol and bornyl acetate/lavandulyl acetate are critical pairs, highlighting the complexity of EOs.

2.4.1.3   Compounds with toxicological background

The group of alkenylbenzenes (Götz *et al.*, 2023) as well as some monoterpenes (Wojtunik-Kulesza, 2022) are under analytical observation due to suspected toxicological effects. Among the alkenylbenzenes, safrole, estragole and β-asarone are of particular importance. Due to their postulated genotoxic potential, the addition of these alkenylbenzenes has already been prohibited in the EU via Regulation (EC) No. 1334/2008. If they are (prominently) present in an EO, as in Myristicae fragrantis aetheroleum (safrole; *Myristica fragrans* Houtt., Myristicaceae), Anisi and Foeniculi amarae fructus aetheroleum (in both latter estragole), they are among the reference compounds for quantification in PhEur (Table 2.1). β-Asarone is present in the EO of *Acarus calamus* L. rhizome. Suspected monoterpenes are thujone and ascaridole. Thujone can be especially found in *Salvia sclarea* L. and *S. lavandifolia* VAHL (both Lamiaceae) EOs, whereas ascaridole can be found in EO of Boldi folium (*Peumus boldus* MOLINA, Monimiaceae).

2.4.2   Generation of chromatographic fingerprints by TLC/HPTLC

Analyses of EOs via TLC or HPTLC regarding identity do not meet the criteria of a chromatographic profiling (implies specific information on particular components) and can be instead counted as fingerprinting (implies unspecific information; Tistaert *et al.*, 2011). PhEur usually selects two or three substances from the EO as reference substances for carrying out the TLC/HPTLC. These compounds can be identified in the EO test solution after the chromatography and serve as anchor points for the R_F values. All other zones of the EO test solution are only identified according to colour and R_F values, but their identity remains unclear.

Although PhEur permits TLC and HPTLC as equally valid methods in the identity section, the introduction of HPTLC represents a major analytical advance. The separation on a smaller and more homogeneous particle size of 2–10 µm in HPTLC has significantly enhanced its reliability in terms of resolution and detection limits, making it more valuable for HPTLC fingerprinting and also chemical profiling studies for plant species authentication, as well as for the evaluation of the consistency and stability of botanical drugs and products (Reich and Schibli, 2011). This is accompanied by shorter running distances (often 6 vs 15 cm) and smaller application volumes (often 2 vs 10 µL) of the (reference) substances, resulting in lower costs.

Recording chromatographic fingerprints via TLC/HPTLC and chromatographic profiles via GC–FID, in combination with the physico-chemical values and the methods of pharmacognosy, should be generally suitable for detection of most less sophisticated impurities and instabilities. The PhEur also includes odour and taste evaluations, which are also very helpful in detecting various adulterations. It can be seen as very advantageous that all methods deal with high effectiveness and relatively little instrumental effort. However, as illustrated by examples in the following chapters, numerous analytical challenges cannot be solved with pharmacopoeia methods, and a review of the more recent literature reveals two fundamental shortcomings of the pharmacopoeial methods in particular: (i) the reliable identification of sophisticated adulterations requires the evaluation of analytical measurements using appropriate chemometric methods which are not sufficiently established in pharmacopoeias, and (ii) the analytical method spectrum is not broad enough in relation to modern coupled instrumental methods.

2.5　EO-Containing Plants in PhEur

Among and besides the ~330 monographs listed in the PhEur chapter
'Herbal Drugs and Preparations of Herbal Drugs' there are also a total of ~45
monographs in which the EO content is determined in the assay section. Thus,
besides the flavonoids and phenolic acids, EOs are one of the most frequently
used groups of secondary metabolites to establish a uniform quantitative
assay for medicinal plants. Interestingly, in Boldo leaves (*Peumus boldus* MOLINA,
Monimiaceae), volumetric determination of the EOs is part of the purity test
and the content of EO is limited upwards and must not exceed 40 ml/kg
(Christoffers *et al.*, 2014).

2.5.1　Extraction

The general aim of the PhEur to make analytical methods as uniform as possible (harmonization) is also reflected in the extraction technique chosen. For
all monographs, hydrodistillation is recommended, using a uniform apparatus
consisting of a round-bottom flask, a standardized steam distillation apparatus, boiling water and an organic solvent for absorption of EO. After setting a
constant distillation speed and volume, the plant material is added and the EO
collected until the end of distillation.

Nevertheless, solid phase microextraction (SPME) has been introduced as
a modern alternative to traditional hydrodistillation (Arthur and Pawliszyn,
1990). This technique eliminates the use of organic solvents and shortens the
time of sample preparation before analysis. In one approach, a partitioning
equilibrium between the sample matrix and the extraction phase is reached.
A second gentle extraction method is the use of supercritical carbon dioxide.

2.5.2　Volumetric quantification

Determination of the EO content is always done after the hydrodistillation and
expressed as ml/kg dried plant material. The EO can be collected alone or in
an organic solvent. As the standard hydrodistillation apparatus contains a
graduated tube and a three-way adapter, the volume of EO can be determined
precisely after draining a suitable amount of water.

Within one monograph, the plant material can be defined as comminuted,
cut or uncut, and consequently PhEur sometimes defines a lower content of
EO for the comminuted or cut form as it loses the EO more quickly due to its
larger surface area.

Also, the definition of an own monograph is possible, as can be observed
for Valerianae radix minutata (from *Valeriana officinalis* L. s.l., Caprifoliaceae), a
special cut material for tea preparation. Interestingly, in exceptional cases, different plants are permitted in one monograph, even if they have a significantly
different content of EOs. This resulted in the definition of different requirements for both plants, as can be seen in the monograph Amomi fructus. For
Amomum villosum LOURDES, the required content of EO is 30 ml/kg, whereas for

the also approved *Amomum longiligulare* T.L.Wu (both Zingiberaceae) only 10 ml/kg is necessary. This differs procedurally from other secondary metabolites groups, where two different monographs are created as can be observed, for example, for Aloe.

Case in point: EO content of uncut vs cut plant material

Content of EO in dried uncut leaves of Eucalypti folium (*Eucalyptus globulus* Labill., Rutaceae) must count for ≥20 ml/kg, and for dried cut leaves only ≥15 ml/ kg. For Levistici radix uncut (*Levisticum officinale* W.D.J. Koch, Apiaceae), the EO content should be ≥4.0 ml/kg, whereas cut material must count for only ≥3.0 ml/ kg. In Menthae piperitae folium (*Mentha* ×*piperita* L., Lamiaceae), uncut material must count for ≥12 ml/kg, whereas cut material should contain ≥9.0 ml/kg, and in the leaves of *Salvia officinalis* L. (Lamiaceae) content limits are ≥12 ml/kg for uncut and ≥10 ml/kg for cut material.

2.5.3   Downstream assay methods

After hydrodistillation and volumetric quantification, the obtained EO can be subjected to further downstream analyses, which can be part of a purity test or assay. Quantification of one or more value-determining compounds is the motivation for using GC–FID as a downstream method in the assay section using respective references and normalization (Section 2.3). In Amomi fructus (*A. villosum and A. longiligulare*), bornyl acetate must be present at ≥30%, whereas Amomi fructus rotundifolius (*A. krervanh* Pierre ex Ganep. and *A. compactum* Sol. ex Maton) must contain ≥65% 1,8-cineol. The content of thymol/carvacrol in Thymi herba (*Thymus vulgaris* L., *T. zygis* L.) must count for ≥40% and the content of *trans*-anethole in Anisi stellati fructus (*Illicium verum* Hook. f., Illiciaceae) must exceed ≥86%.

Quantification via GC–FID can also be a part of the purity test. It concerns mainly marker compounds for differentiation of two medicinal plants or substances that are suspected to be toxicologically problematic. In Foeniculi amari fructus (from *Foeniculum vulgare* Mill. ssp. *vulgare* var. *vulgare*), the content of estragole is limited to 5%, and in Foeniculi dulci fructus (from *F. vulgare* ssp. *vulgare* var. *dulce* (Mill.) Batt. & Trab., both Apiaceae) to 10% estragole as *in vitro* tests revealed that it is bioactivated in metabolism via hydroxylation to form probably genotoxic DNA adducts (de Vincenzi *et al.*, 2000; Schulte-Hubbert *et al.*, 2020). Nevertheless, the impact for humans is still questionable and under discussion (Levorato *et al.*, 2018). Furthermore, in Foeniculi dulci fructus the fenchone content is determined to be ≤7.5% in the purity section for exclusion of contamination with bitter fennel. In contrast, the same compound is a quality-determining substance in Foeniculi amari fructus as it is the bitter principle of the drug and thus, together with anethole (≥ 60%), part of the assay (≥15% fenchone).

Besides the EO, the aerial parts of *Achillea millefolium* (Millefolia herba) accumulate sesquiterpene lactones of the guaianolide type. Depending on their structure, some of them (so-called proazulenes) are decomposed during the hydrodistillation process to form the intensively blue-coloured chamazulene. Due to the anti-inflammatory activity of chamazulene (Ramadan *et al.*, 2006; Flemming *et al.*, 2015) proazulenes are considered value-determining compounds, and the content of chamazulene in EO of Millefolia herba is therefore determined via UV-spectrometry (absorption at 608 nm).

2.6   Analytical Methods Beyond the Pharmacopoeias

2.6.1   Conventional techniques

It seems at a first glance that further highly sophisticated methods to guarantee pharmaceutical quality are not required as standard methods in PhEur because highly sophisticated adulterations are not commonplace in these EOs. This consideration would follow the general rule that requirements in monographs are based, for example, on previous experiences of process values and adulterations occurring in the markets. Hitherto unknown adulterations first must find their way into praxis and literature before their detection will be included in a pharmacopoeia. Nevertheless, analytical research on EOs beyond the pharmacopoeias is extremely diverse and motivated by several market- and scientific-driven reasons.

An in-depth look at the list of monographs with EOs in PhEur clearly shows that many of the most expensive EOs are not included. These are (and due to strongly fluctuating world market prices, without claiming to be exhaustive) elecampane EO (from flowers of *Inula helenium* L., Asteraceae), sandalwood EO (from the heartwood of *Santalum album* L. and other *Santalum* spec., Santalaceae), seaweed absolute oil (from various macroalgae, e.g. *Fucus vesiculosus* L., Fucaceae), jasmine absolute oil (from the blossoms of *Jasminum grandiflorum* L. and *J. sambac* (L.) Aiton, Oleaceae), agarwood oil (oud attar, from the wood of *Aquilaria malaccensis* Lam., Thymelaeaceae), hemp oil (cannabis flower EO, from flowers and upper leaves of *Cannabis sativa* L., Cannabaceae), frangipani absolute EO (from flowers of *Plumeria alba* L., Apocynaceae), tuberose absolute EO (from flowers of *Agave polianthes* Thiede & Eggli, Asparagaceae), rose otto EO (from flowers of *Rosa × damascena* Mill., Rosaceae), champaca absolute EO (from flowers of *Magnolia champaca* (L.) Baill. ex Pierre, Magnoliaceae) and iris EO (from the roots of *Iris germanica* var. *florentina*, Iridaceae). Thus, many of the most expensive EOs are used as fragrances in the cosmetics industry and/or in aromatherapy. The prices of these EOs are 500–2500 (and more) Euro/ounce (~28 g), so that even more elaborated adulterations are 'worthwhile' and commonplace.

Case in point: Adulterations of EOs

As an example, several adulterations of sandalwood EO with synthetic alcohols, such as javanol, polysantol and ebanol, or with *Amyris balsamifera* L. ('West Indian sandal wood', Rutaceae) EO were reported and none of the investigated samples met the ISO criteria of sandalwood EO (Kucharska *et al.*, 2021). Recently, various adulteration techniques for *Rosa × damascena* EO were reported using individual (natural or synthetic) substitutes such as phenylethanol, citronellol, geraniol, geranyl acetate and linalool or mixing rose oil with cheaper EOs such as palmarosa oil (*Cymbopogon martini* (Roxb.) Wats., Poaceae) or non-volatile components such as vegetable oils (Raeber *et al.*, 2023).

It likely can be assumed, the more expensive the traded EO, the more advanced the adulteration and the method needed for detection. Interestingly, many of these expensive EOs also differ to some extent from the 'Aetherolia' of the pharmacopoeia in terms of the extraction technique, as they include numerous 'concretes' and 'absolutes', i.e. oils that are not obtained by hydrodistillation, but by solvent extraction at room temperature or lower. As an example, production of jasmine absolute from flowers involves a two-step method where hexane is first used to extract 'jasmine concrete' (a solid, waxy-buttery extract consisting of oil and wax). The concrete can be commercialized as it is or converted into 'jasmine absolute', a fragrance material free of waxes. This is performed by washing the concrete numerous times with ethanol (at minus temperatures) to allow the soluble fraction on the concrete to dissolve into the ethanol, while preventing the flaking wax from doing so and removing it by filtration.

Besides their use in fragrances and aromatherapy, several EOs are obtained from food plants, which are also not regulated by the pharmacopoeias. A typical example is *Vanilla planifolia* Andr. (Orchidaceae), providing an EO as well as the aroma compound vanillin, the latter being needed on a scale of more than 12,000t per year (Straits Research, 2021). Furthermore, some plants are used on a large scale beyond their medical applications as spices or food ingredients (Chapter 7), such as cinnamon bark (*Cinnamomum verum* J.Presl., Lauraceae), lemon balm leaves (*Melissa officinalis* L.) and peppermint leaves (*Mentha ×piperita*, both Lamiaceae). Here, the high tonnages required are an attractive target for adulterations. Consequently, a complex example for various analytical needs, including markers and chromatographic profiles, is peppermint oil (*Mentha ×piperita* L., Lamiaceae) as it is often adulterated with cheaper EOs or single compounds from closely related *Mentha* species like spearmint (*Mentha spicata* L.), menthol mint (*Mentha arvensis* L., corn mint oil) and pennyroyal (*Mentha pulegium* L., pennyroyal oil, all Lamiaceae). The identity of peppermint oil can be determined by the markers *trans*-sabinene at the 1% level or viridiflorol, which are present only in *M. piperita* (Singhal *et al.*, 1997). Quantitatively larger adulteration with *M. spicata* oil can be detected by the calculation of the menthone/isomenthone ratio that must be between 5.0 and 9.2 (Pruthi, 1998, p. 459).

Accordingly, there is an enormous technical and methodological need for quality assurance of high-quality, unadulterated phytogenic EOs in cosmetics, aromatherapy and foods beyond the pharmacopoeias. In the context of highly sophisticated adulterations, three main analytical challenges have emerged:

1. Detection of cheaper, but very similar natural EOs in smaller amounts.
2. Separation and differential detection of racemates to test the addition of a racemic compound to an EO containing phytogenic substances with high enantiomeric purity.
3. Detection of chiral synthetic, but enantiomerically pure, nature-identical impurities.

For (1) modern analytical (hyphenated) methods have recently been moving in the direction of a combination of chromatographic and spectroscopic profiling, which is combined with multivariate statistical analysis. As examples, Fourier transform infrared spectroscopy, Raman or nuclear magnetic resonance spectroscopy as well as different mass spectrometry techniques have found their way into the analytics of EOs.

Point (2) is addressed in pharmacopoeias, as PhEur accepts the challenge to test the addition of a racemic compound to an EO with high enantiomeric purity, but only in a few monographs by using an enantioselective chromatography with chiral phase β-cyclodextrin.

For (3) the detection of chiral synthetic, but enantiomerically pure, nature-identical impurities is addressed by carbon isotope ratio analyses. These analytical methods make use of the biosynthetic isotope discrimination of the plant in comparison to synthetics. Suitable techniques for authentication of EOs include isotope ratio mass chromatography (IRMS), which can be coupled to GC as GC–IRMS. The type of column used is generic, whereas the MS detector is not. There are two factors influencing isotopic ratios, with one being the addition of synthetic components that were made from fossil fuel-derived scaffolds. In this case, the natural process of radioactive decay substantially reduces the amount of ^{13}C in fossil fuels over time, which reduces the $^{13}C/^{12}C$ ratio. For the second, it is more common for IRMS to be applied where multiple EOs of natural origin are combined to falsify a more expensive EO. As plants utilized different forms of CO_2 fixation (C3-, C4- and CAM-plants), the difference in isotopic ratios results from photosynthesis. Plants that follow C3 photosynthesis (Calvin cycle) such as *Citrus* spec. do not accumulate as much ^{13}C as C4-plants (Hatch–Slack cycle) such as *Cymbopogon* spec. When lemon essential oil from a C3 plant (*Citrus* spec.) is adulterated with citral from a C4 grass species (*Cymbopogon* spec), the isotopic ratio is biased towards higher ^{13}C amounts. Plants following CO_2 fixation of the Crassulacean type (CAM-plants), create an isotopic ratio in between that of C3 and C4 plants. As an example, the determination of enantiomeric and stable isotope ratio fingerprints of active secondary metabolites in neroli essential oil for authentication by multidimensional gas chromatography and gas chromatography–combustion/pyrolysis–isotope ratio mass spectrometry has been successfully implemented (Cuchet *et al.*, 2021). Also, for authentication of natural vanillin, a GC-IRMS method has been established (Hansen *et al.*, 2014).

Another spectroscopic concept of isotope discrimination is site-specific natural isotope fractionation-NMR (SNIF-NMR). This technique uses ^{2}H NMR spectroscopy and is able to measure non-statistical distribution of deuterium in different sites of a given molecule. Thus, it can be utilized to differentiate between its natural and synthetic origin. For example vanillin can thus be analysed whether it is natural, synthetized from guaiacol, synthesized from lignin or produced biotechnologically and can be easily combined with ^{13}C NMR techniques (Guyader *et al.*, 2019).

Besides the 'economically driven' analytical research, there is also a 'scientifically driven' need for research on EOs aiming to avoid scientific mistakes or to improve an analytical procedure, technically and methodically. As an example, decomposition of labile substances in EOs is a general analytical challenge and touches on all steps of extraction, processing and sampling (Mahanta *et al.*, 2021). Numerous new analytical approaches have been established in these fields, which have not yet been reflected in PhEur. For example, the application of HPLC for analysis of EOs is a current topic and particularly useful for heat-labile compounds (converted by evaporation of the EO in the GC oven; Turek and Stintzing, 2011). In addition to the improved possibility of analysing heat-labile substances, HPLC has the advantage that other (enantioselective) detectors can be used more conveniently than in GC, e.g. chiroptical coupling. As an example, Lafhal *et al.* (2020) described the creation of a chiroptical fingerprint of lavender using HPLC-polarimetric detection. HPLC is also better suited for the simultaneous detection of other substance groups that cannot be vaporized without decomposition together with the EO.

Worth mentioning also is the introduction of ^{1}H NMR in the analyses of thermolabile EO substances (Mahanta *et al.*, 2020).

The most frequently used option for analysis of heat-labile compounds in EOs is the use of more gentle probe application methods in GC, such as different types of dynamic and static headspace techniques (Rubiolo *et al.*, 2010; Louw, 2021). Headspace is a common solvent-free method aiming at sampling the gaseous or vapour phase in equilibrium with a solid or liquid matrix. This is to avoid the use of organic solvents as well as thermal-based extractions. Quite a large number of conventional techniques may be used to sample the volatile fraction: vacuum-, steam- or hydrodistillation, solvent extraction off-line combined with distillation, simultaneous distillation-extraction (SDE), supercritical fluid extraction (SFE) and microwave-assisted extraction and hydrodistillation (MAE and MA-HD). Nevertheless, it should be stated that, per international definition, EOs are products obtained by hydro-, steam- or dry-distillation or by a suitable mechanical process without heating (for

Citrus fruits) of a plant or of some parts of it. Consequently, substance blends obtained from headspace (and some other extraction procedures) are volatile mixtures, but not EOs anymore. In PhEur, the chosen sample application for GC–FID, a liquid injection into a heated injector with adequate split ratio, is the universal standard sampling method.

An analysis of recent literature on EOs makes clear that for several years the coupling of GC with a mass-selective detector has been the most common and powerful hyphenation (GC-MS). It combines structural information with high sensitivity and is extremely powerful for the identification and chromatographic profiling of EOs when using a database with relative retention times, fragmentation pattern and robust data analysis procedures (Lebanov *et al.*, 2021; Sadgrove *et al.*, 2022). The combination of mass spectrometry with comprehensive and multidimensional chromatography (such as GC ×GC and LC-GC) and chemometry can cope with challenges of chemotyping, enantioseparation, quality control and adulteration of EOs (Rasheed *et al.*, 2021).

At first glance, it seems astonishing that GC-MS coupling cannot be found in PhEur at all. Nevertheless, it becomes clearer if one analyses the objectives of the pharmacopoeias in more detail. The objective of the pharmacopoeia is to set routine standards for the quality of medicinal products, and these standards should be developed in such a way that they also meet the needs of the different persons responsible for quality control in industry and pharmacy.

2.6.2 Effect-directed technologies

On the one hand, the established target methods and pharmacopoeial monographs focus on known substances or substance groups to detect or quantify already known markers, adulterants, residues, impurities or contaminants in EOs. On the other hand, comprehensive methods based on multidimensional column chromatography with different separation and detection principles result in thousands of signals that can be detected. However, only a minor part can be assigned to known substances. Most signals remain of unknown identity and unknown toxicity. Important compounds can also be overlooked, as not all molecules may be ionized equally well, and some may not be ionized at all using default settings of mainstream mass spectrometric detection. Such a comprehensive analysis involves considerable effort and expense and is not applicable in routine control. It is tempting to set an intensity threshold or to focus on very frequent signals. However, this can mislead decisions as even a small signal may represent a compound with an important effect.

Given the complexity of EOs, there is a need for non-target effect-directed analytical techniques with a prioritization strategy (Fig. 2.2) that can evaluate and control the quality of such complex EO mixtures more comprehensively regarding active compounds present (Morlock, 2022). Combining planar chromatography with an effect-directed assay is an emerging technique, which prioritizes active compounds among the thousands of compounds present in such complex mixtures (Móricz *et al.*, 2015; Teh and

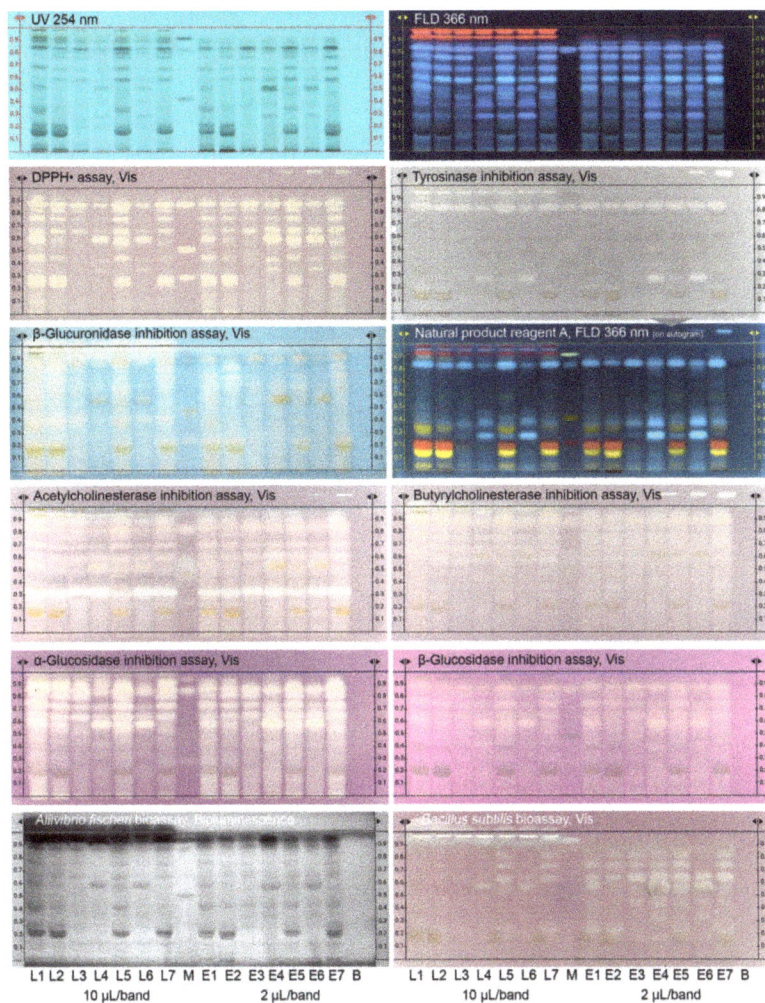

Fig. 2.2. Bioprofiling of peppermint products using different bioactivity mechanisms: HPTLC–Vis/UV/FLD chromatograms of the various leaf samples L1–L7 and respective powdered extracts E1–E7 along with standard mixture (M; eriocitrin, luteolin-7-O-glucoside, rosmarinic acid, and apigenin, 1.5 µg/band each) and the respective solvent blank (B) separated on HPTLC plate silica gel 60 F$_{254}$ s with 7 ml ethyl acetate–toluene–formic acid–water 8:2:1.5:1 and detected at UV 254 nm, FLD 366 nm and after the respective (bio)assay at white light illumination (bioluminescence depicted as greyscale image for *A. fischeri*). Derivatization with natural product reagent A was performed after the assay on the tyrosinase inhibition autogram. (Inarejos-Garcia *et al.*, 2023; CC BY.)

Morlock, 2015; Bañuelos-Hernández *et al.*, 2020). Not only known but also unknown bioactive compounds are detected. The detection of both allows for identifying important bioactive natural or process-related changes, adulterants, residues and contaminants (Morlock and Meyer, 2023; Müller *et al.*, 2023). Thereby, the matrix-robust technique with a minimalist sample preparation is advantageous for sample integrity, as components can be lost or discriminated at each step of sample preparation. Effect-directed profiling via the combination of chromatography with effect-directed assays provides important information for compound prioritization and should be applied more frequently to EOs.

If a previously unknown but active and therefore relevant signal needs to be identified, the given possibility and flexibility to hyphenate different separations (elution of an active zone via an orthogonal column separation into the MS, and thus coupling normal phase separation and reversed phase separation) and detection principles is favourable. In the same chromatographic run, the sample can be detected by UV/Vis/FLD, diode array detection (DAD), high-resolution mass spectrometry with fragmentation capability (HRMS/MS) and derivatization sequence. Multiple detection using different detection principles is advantageous to capture information on the variety of compounds as comprehensively as possible. In particular, HRMS/MS is helpful to tentatively assign the newly discovered, previously unknown active compounds (Schreiner *et al.*, 2021; Ronzheimer *et al.*, 2022). Such an effect-directed profiling can be applied in routine analysis, as it is fast, robust and low-cost.

The recently introduced open-source 2LabsToGo-Eco (Sing *et al.*, 2022; Romero *et al.*, 2025) provided results comparable to those of the status quo instruments but at significantly lower equipment and laboratory infrastructure costs (Fig. 2.3), beneficial in situations where laboratory resources are scarce (Morlock, 2022; Morlock *et al.*, 2023; Morlock and Heil, 2025). The portable all-in-one 2LabsToGo-Eco strikes the perfect balance in miniaturization and ease of handling, not too small for human hands. Its very low instrumental footprint (dematerialization) and the straightforward prioritization strategy help analytical chemistry to balance between technology and nature/ecology to reduce the planetary overshoot. The technology can be used to reach a broader market coverage through increased low-cost screening as well as quantification capabilities and to filter out among the thousands of compound signals those that show either beneficial or harmful biological (toxicological) activity (Fig. 2.4).

2.7   Chemometrics

The term 'chemometrics' refers to mathematical and statistical methods that are required to process and analyse large amounts of measured data in a suitable manner, e.g. in the context of an analytical (e.g. spectroscopic, spectrometric or chromatographic) or (more general) chemical process. Typical

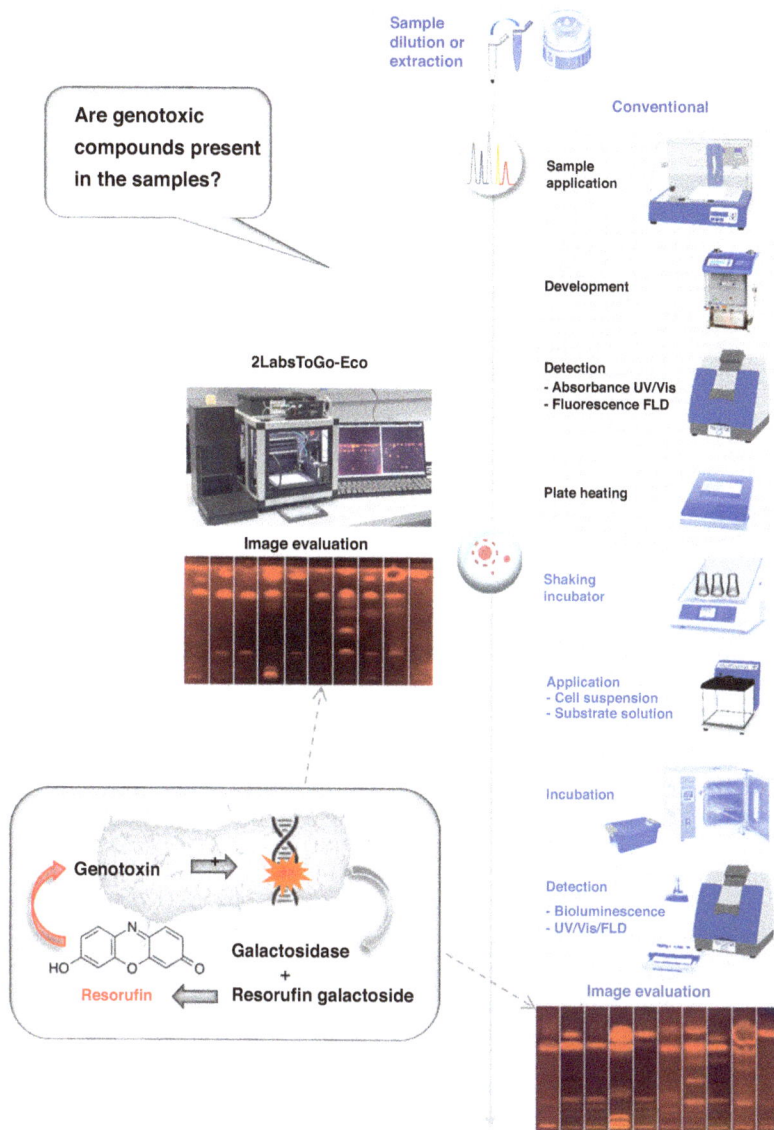

Fig. 2.3. Schematic comparison of two different instrumental approaches for effect-directed analysis, exemplarily depicted for screening of genotoxic compounds in complex mixtures, detected as orange fluorescent resorufin zones. (Adjusted from Morlock *et al.*, 2023; Morlock and Heil, 2025. CC BY-ND 3.0.)

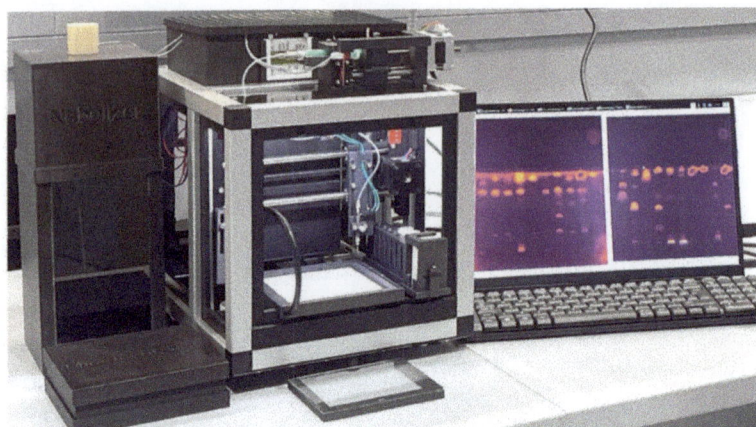

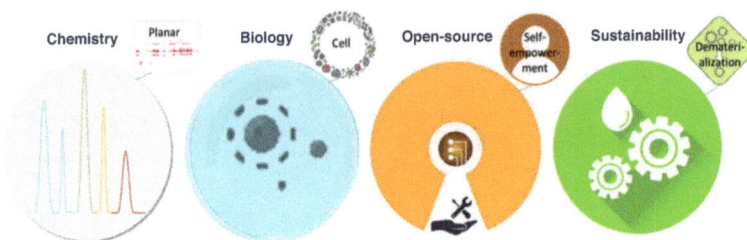

Fig. 2.4. Sustainable open-source 2LabsToGo-Eco consolidating the functionalities and operations of all devices in two laboratories (HPTLC lab and bioassay lab) in a miniaturized system; it combines different fields and technologies to allow for effect-directed analysis via planar bioassays. (Adjusted from Morlock and Heil, 2025; CC BY-ND-3.0.)

standard procedures are principal component analysis (PCA), often comprising the first step of data interpretation, recognition of sample grouping, trends and strong outliers. In a second step, application of hierarchal cluster analysis (HCA) can be carried out for clustering measured data. Further methods include, among others, partial least square discriminant analysis (PLS-DA) to find the fundamental relations between two matrices.

It is obvious that chemometric methods are very helpful in analysing EOs, as they are per se complex mixtures of substances whose complexity is further increased by the fact that they can be adulterated by other natural or synthetic compounds or EOs. Several analytical methods on EOs are either coupled (e.g. GC-MS) or even multidimensional (e.g. GC × GC-MS) so that the increasing data volumes have to be analysed mathematically and statistically. Numerous application examples for chemometric analyses of coupled or uncoupled methods can be found in literature as follows.

Chemometrics provides important information when analysing essential oils from the same species but different origins (Sahoo *et al.*, 2022) or in chemical profiling of different species (Clery *et al.*, 2022). Chemometrics can help to identify different types of adulterations (Cebi *et al.*, 2020; Truzzi *et al.*, 2021) or for authentication (Huang *et al.*, 2022). Last but not least, pharmacological data can also be included in the chemometric analysis (Youssef *et al.*, 2020), and, accordingly, in current analytical or analytical-pharmacological studies on essential oils one can hardly find a publication without chemometric procedures.

Galenic Aspects – Processing and General Dosage Forms of Essential Oils as Multi-component Mixtures

Rolf Daniels*

3.1 Definition of Essential Oils

European Pharmacopoeia (PhEur) defines an essential oil (EO) as 'odorous product, usually of complex composition, obtained from a botanically defined herbal drug. These herbal drugs may be fresh, lightly wilted, wilted, partially dried, dried, whole, fragmented, broken or cut. The herbal drug may also undergo a preliminary treatment. Processes which are used to obtain EO are steam distillation, dry distillation, or a suitable mechanical process without heating' (EDQM, 2024).

This definition is well in line with the definition by the ISO in document ISO 9235:2021 on 'Aromatic natural raw materials' which states the following: 'EO: product obtained from a natural raw material of plant origin, by steam distillation, by mechanical processes from the epicarp of citrus fruits or by dry distillation, after separation of the aqueous phase – if any – by physical processes' (International Organization for Standardization, 2021).

Furthermore, ISO 9235 states that 'The EO can undergo physical treatments which do not result in any significant change in its composition (e.g. filtration, decantation, centrifugation)' (International Organization for Standardization, 2021).

This is in principle also in line with PhEur. However, according to PhEur, it is possible to subject EOs to further processing steps (combining, rectification, etc.), that may or may not significantly affect its composition. In order to make such changes transparent, the type of modification must be indicated in the Definition section of an individual monograph. Typical terms used in

*Corresponding author: rolf.daniels@uni-tuebingen.de

this context are: rectified EO, deterpenated and desesquiterpenated EO (EDQM, 2024).

Independent from these differences dealing with post-treatment, it is common understanding that EOs are not defined by their composition but by their origin and the production process.

As a result, products obtained through alternative extraction methods, such as solvent extracts (including supercritical carbon dioxide extracts), concretes, pomades, absolutes, resinoids and oleoresins, do not strictly adhere to the common definition of EOs.

3.2 Essential Oil Production

There are three methods employed to extract EOs from plant material (Schmidt, 2016).

1. The first method, expression, predominantly utilized for extracting EOs from the peel of citrus fruits, has the longest history. It refers to physical processes in which the EO glands in the peel of citrus fruits are crushed or broken to release the oil (Phi *et al.*, 2015).
2. The second method, hydrodistillation or steam distillation (Fig. 3.1), stands as the most widely employed technique among the three, also with a long history (Levey, 1959; Rovesti, 1977; Schmidt, 2016).
3. Conversely, dry distillation, as the third method, finds rare application, typically reserved for specific cases such as the production of cade oil from the wood of *Juniperus oxycedrus* L. (Schmidt, 2016).

The most common and very simple method to produce EOs is by means of hydrodistillation or steam distillation (Masango, 2005). Important for this method is that the substances to be isolated should be almost insoluble in water. This is mostly the case with EOs. In ideal mixtures that are not miscible or partially soluble, a total vapour pressure is established, which is the sum of the partial vapour pressures of the individual components.

$$\text{Total vapour pressure } p_{total} = \text{Partial pressure } p_{water} + \text{Partial pressure } p_{essential\ oil}$$

If the vapour pressure is higher than the external pressure, a liquid begins to boil. In mixtures that are immiscible, this increases the total vapour pressure p_{total}, which is composed of the partial pressures p_{water} and $p_{essential\ oil}$. Consequently, this results in a lowering of the boiling temperature, which is therefore always below 100°C, guaranteeing mild isolation conditions. This is especially important for heat-sensitive natural substances such as EOs.

During boiling, the water vapour carries the heat-sensitive compounds of the EO with it. That is why this process is also referred to as steam distillation. Upon cooling, the aqueous phase separates from the insoluble components, allowing the EO to be obtained.

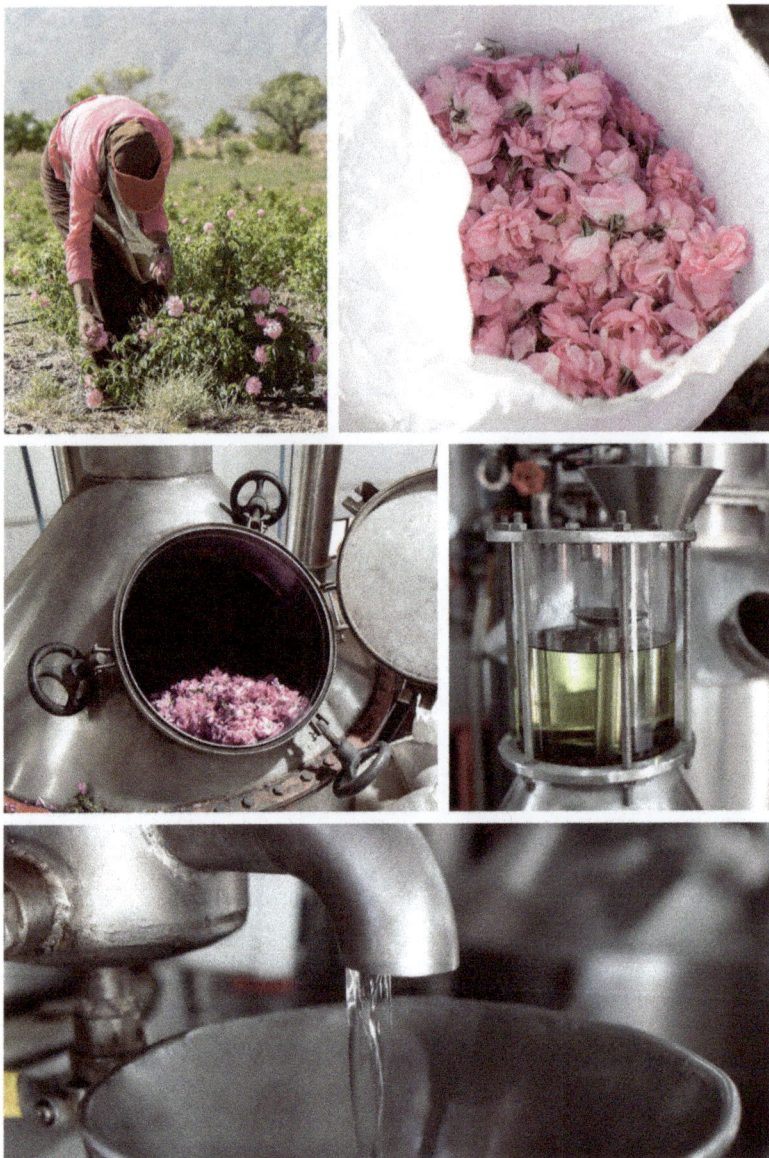

Fig. 3.1. Production of essential oil from Damask roses (*Rosa × damascena*).
(Photos: Courtesy of WALA Heilmittel GmbH.)

The volatility of the oil constituents is not influenced by the rate of vaporization but does depend on the degree of their solubility in water. As a result, the more water-soluble essential components will distil over before the more volatile but less water-soluble ones.

In addition to the EO, the so-called hydrolate is obtained. This is the separated aqueous phase which is capable of dissolving small amounts of more polar compounds of the EO.

After the primary extraction process, some EOs are rectified. This involves redistillation of the crude oil to remove certain undesirable impurities, such as trace amounts of components of either very low or high volatility which reduce the EO's quality, e.g. by producing an off-odour, or toxic compounds such as methyleugenol from rose oil. During rectification, overheating followed by partial decomposition of susceptible constituents has to be avoided. Therefore, it is frequently performed by redistillation under vacuum or by steaming (Schmidt, 2016).

3.3 Formulation concepts

Essential oils are formulated into several dosage forms depending on the intended route of administration (Table 3.1).

From a formulator's point of view, three major properties of EOs are to be considered. Essential oils are (i) *volatile*, (ii) *fluid* and (iii) *lipophilic*. Mixing EOs with suitable liquid lipids can easily be done. Such oily solutions can be applied to the skin or can be used as filling material for soft capsules, which are usually administered orally (Gullapalli, 2010). Filling into hard capsules for oral use can also be an option. However, hard capsules can only be used when respective measures are taken to avoid leaking of the oily content, e.g. by banding (Cadé *et al.*, 1986).

The formulation into liquid or semi-solid emulsions can also be easily done as the EO mixes with the oil phase of the emulsion. Due to their partially amphiphilic character, EOs may interact with the emulsifier system and subsequently destabilize emulsions. Consequently, a proper selection of the emulsifiers used is mandatory (Septiyanti *et al.*, 2017; Martin *et al.*, 2018a).

Aqueous solutions require either acceptable cosolvents, e.g. sufficiently high amounts of ethanol, or the addition of hydrophilic surfactants as solubilizing agents, e.g. polysorbate 80, which can accommodate the EO in their lipophilic core conveying the formulation into a clear solution. Alternatively, EO can be molecularly encapsulated in the hydrophobic interior of cyclodextrin molecules, also yielding clear aqueous solutions (Marques, 2010; Salústio *et al.*, 2015). All kind of aqueous solutions can be applied to the lining of the mouth. They can also be inhaled or used as bath additives.

Further liquid formulation concepts comprise nano-particular systems like micro- and nano-emulsions (Pavoni *et al.*, 2020; Thakur *et al.*, 2021;

Table 3.1. Essential oils covered by Committee on Herbal Medicinal Products (HMPC) monographs and their corresponding dosage forms. (From European Medicines Agency, undated. Author's own table using publicly accessible data.)

Essential oil	Peroral					Dermal			Lining of the mouth	Liquid dosage forms for dental and oromucosal
	Solid	Semi-solid	Liquid	Tea	Liquid inhalation	Bath additives	Liquid	Semi-solid	Liquid forms	
Anise oil	x		x							
Bitter fennel fruit oil				x						
Caraway oil			x					x		
Cinnamon bark oil			x							
Clove oil										x
Eucalyptus oil	x		x		x	x	x	x		
Juniper oil			x				x			
Lavender oil			x			x				
Matricaria oil						x				
Peppermint oil	x	x	x		x		x	x	x	
Rosemary oil								x		
Tea tree oil							x	x	x	
Thyme oil			x			x	x	x		
Valerian essential oil			x			x				

Pandey *et al.*, 2022), Pickering emulsions (Souza *et al.*, 2020), solid-lipid nanoparticles (Cimino *et al.*, 2021), nanostructured lipid carriers (Ghodrati *et al.*, 2019; Baldim *et al.*, 2022) and liposomes (Sherry *et al.*, 2013; Hammoud *et al.*, 2019). Most of these formulation concepts are intended to enhance the bioavailability and allow for controlled delivery of the EO but also to improve stability by protection from oxygen exposure (Saifullah *et al.*, 2019; Weisany *et al.*, 2022; Yammine *et al.*, 2024).

The most demanding formulation option is the transformation of an EO into a dry powder or granules. This can be done by several encapsulation techniques which all aim to produce a solid shell protecting the liquid core consisting of the EO (Majeed *et al.*, 2015; Avila Gandra *et al.*, 2018; Lammari *et al.*, 2021; Mukurumbira *et al.*, 2022). The wall material can be chosen from numerous chemical classes, for example, gums (Daniels and Mittermaier, 1995; Martins *et al.*, 2017), natural or modified polysaccharides (Dierings de Souza *et al.*, 2021; Shlosman *et al.*, 2022; Lim *et al.*, 2023), chitosan (Javid *et al.*, 2014), lipids, proteins (Xue *et al.*, 2019) and diverse others (Avila Gandra *et al.*, 2018).

Numerous techniques are described for micro- and nano-encapsulation of EOs (Majeed *et al.*, 2015; Avila Gandra *et al.*, 2018; Lammari *et al.*, 2021; Mukurumbira *et al.*, 2022; Weisany *et al.*, 2022). Among the methodologies described for the encapsulation are spray-drying, extrusion, fluidized bed coating, coacervation and ionic gelation. Electro-spinning and electro-spraying are advanced techniques which are employed to produce solid material with encapsulated EO (Cengiz Çallıoğlu and Kesici Güler, 2020; Pires *et al.*, 2023). Recent studies also focused on micro- and mesoporous materials for encapsulation and controlled release of EOs with the advantage that they require less carrier excipients and have a superior loading capacity (Su *et al.*, 2024).

Apart from all these innovative formulation concepts, spray-drying is still the most frequently used method for encapsulating EOs owing to its capacity for large-scale continuous production, cost efficiency and flexibility in select-ing appropriate coating materials, including carbohydrates and proteins, pro-viding high flexibility in formulation design (Reineccius, 2004; Gharsallaoui *et al.*, 2007; Da Veiga *et al.*, 2019; Altay *et al.*, 2024; Phanse and Chandra, 2024). Spray-drying can be described as a four-step procedure: (i) preparation of a homogeneous oil-in-water emulsion of the EO in the aqueous solution of the wall material, which frequently comprises a polymer, (ii) atomization of the emulsion by means of a nozzle or a rotary atomizer, (iii) moisture evapora-tion using mainly hot air or less frequently nitrogen gas, and (iv) separation of the final product from the drying air. During drying of the dispersion, rarely singular droplets of the EO are encapsulated. Mostly matrix systems are formed where EO droplets are incorporated into a dry excipient matrix separating the EO from the environment.

Extrusion is another technique to embed EO in matrices. Extrusion involves forcing a mixture of EO and a carrier material through a die under

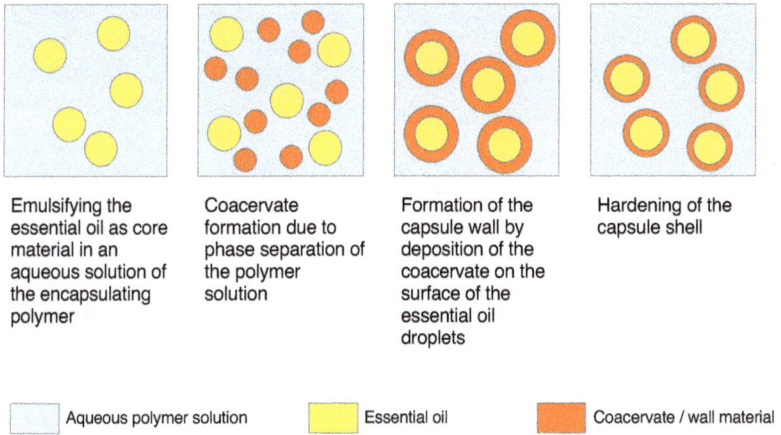

Emulsifying the essential oil as core material in an aqueous solution of the encapsulating polymer	Coacervate formation due to phase separation of the polymer solution	Formation of the capsule wall by deposition of the coacervate on the surface of the essential oil droplets	Hardening of the capsule shell

☐ Aqueous polymer solution   ▮ Essential oil   ▮ Coacervate / wall material

Fig. 3.2. Coacervation microencapsulation as a four-step process. (By Rolf Daniels.)

high pressure to form uniform extrudates or pellets (Azevedo *et al.*, 2019). The carrier material, often a matrix-forming polymer or lipid, serves as the shell material encapsulating the EO core. Like spray-drying, extrusion can be easily scaled up to production scale.

When mostly singular droplets that are covered by a distinct shell are required, simple or complex coacervation techniques may be used (Daniels and Mittermaier, 1995; Timilsena *et al.*, 2019; Muhoza *et al.*, 2022), yielding small-sized micro-capsules. This process (Fig. 3.2) generally involves four steps: (i) the EO as core material is emulsified in an aqueous solution of the encapsulating polymer, (ii) phase separation of the polymer solution, resulting in the formation of a coacervate phase, (iii) formation of the shell wall by deposition of the coacervate on the surface of the core material, and (iv) hardening of the shell, e.g. by cross-linking the polymers. Thereafter, the micro-capsules have to be separated from the aqueous phase and dried to remove any residual solvent and obtain the final micro-capsules as solid material.

Independent from the methodology actually used, covering EO droplets with any wall material enhances their storage stability by protecting them from the external environment by slowing down volatilization and oxidation.

Encapsulated EOs can be used alone or combined with other solids administered orally as powder or granules or filled into capsules. Encapsulated EOs can be added to herbal teas (EDQM, 2023). This is also a common technique when it comes to instant herbal teas (EDQM, 2023). Such powders or granules often comprise EOs which are either added as encapsulated dry powders or are encapsulated *in situ* when the drying of the herbal drug preparation is performed in the presence of suitable excipients such as maltodextrin.

Use of Essential Oil Preparations in Veterinary Medicine

4

Sandra Graf-Schiller, Cäcilia Brendieck-Worm and Matthias F. Melzig*

4.1 Essential Oils and Essential Oil Compounds as Antimicrobials

The increasing prevalence of drug resistance in animal production due to the inappropriate use of antibiotics and antiparasitics in human and veterinary medicine is a serious problem worldwide today. Essential oils (EOs) and their isolated compounds can be used as alternatives to synthetic antimicrobials and antiparasitics in treatment of animals and, consequently, to avoid resistance (Matté *et al.*, 2023).

For the scientific justification of such therapeutic use, evolutionary aspects of the function of EOs must be considered. Their role in the defence against microorganisms and other predators is undisputed (Holopainen, 2004; Arimura *et al.*, 2005; Buriani *et al.*, 2020).

This implies a toxicity with regard to the addressees, i.e. bacteria, fungi, viruses or insects, because without this damaging influence, EOs would not be able to fulfil their tasks. The points of attack of the target organisms are mostly unspecific and, in the case of microorganisms, mainly affect membrane permeability or intracellular membrane structures, the functional impairment of which leads to damage or death of the organisms (Nazzaro *et al.*, 2013; Álvarez-Martínez *et al.*, 2021).

Repair of the lesions is only possible to a limited extent in prokaryotes. This explains many of the antimicrobial activities of EOs. The insecticidal activity of EOs is also largely based on unspecific membrane effects. Repellent effects, on the other hand, are based on interactions with olfactory receptors (Ulland *et al.*, 2008; Oladipupo *et al.*, 2022). In parasites, anticoccidial effects, reduced motility, growth inhibition and morphologic changes have been reported (Evangelista *et al.*, 2022; Matté *et al.*, 2023).

*Corresponding author: matthias.melzig@fu-berlin.de

Higher organisms, such as mammals or humans, have the ability to tolerate or repair damage to a certain extent or to metabolize the damaging agents, i.e. to render them harmless and/or excretable. Differences in membrane structure and the complexity of a multicellular organism with functional compartmentalization and its various barriers also reduce the toxicity of non-specific chemical compounds such as those found in EOs. However, this does not apply to all components of EOs and therefore these substances also have a relevant toxicity that at least restricts the therapeutic use of certain EOs in humans and animals.

4.2   Essential Oils as Alternatives to Antibiotics

The use of EOs or preparations containing EOs as a substitute for antibiotics is widely described, both for animal production and for the treatment of pets. The plants with EOs shown to be active against bacterial strains of animal origin are more or less the same as used in human herbal medicine.

Some studies have shown that EOs are good therapeutic alternatives to antibiotics, to fight against local infections, for example in chronic wounds, and against systemic infections, particularly in intensive farming systems where the microbial pressure is often higher than in extensive farming systems (Dupuis *et al.*, 2022).

In recent years, studies with nano-encapsulated EOs have shown a particularly strong effect, as they guarantee improved bioavailability (Linh *et al.*, 2022).

An up-to-date and detailed overview of the use of EOs as anti-infective therapeutics with regard to pathogenic microorganisms relevant in veterinary medicine can be found in Ebani and Mancianti (2020).

Since a large proportion of the antibiotics produced are used in animal production, which is also a source of the development of antibiotic resistance, the potential of EO in cattle farming will be discussed below as an example. EOs in livestock are used for prevention and treatment of microbial infection and parasites as well as to enhance milk production, animal performance and rumen function.

4.2.1   Mastitis: A serious disease for cattle

Mastitis is the most expensive and prevalent disease in dairy cattle worldwide (Castelani *et al.*, 2019; Lopes *et al.*, 2020), showing negative impact on both animal welfare and food security. In recent years, several studies have indicated that mastitis pathogens are becoming resistant to antimicrobials and the cure rates of conventional treatments are low. Besides contagious pathogens, e.g. non-*aureus Staphylococcus*, *Corynebacterium bovis*, *Mycoplasma* spp., *Staphylococcus aureus*, *Streptococcus agalactiae*, and *Streptococcus dysgalactiae*, environmental pathogens are also triggers of the disease, e.g. *Citrobacter* spp.,

Enterobacter spp., *Escherichia coli*, *Klebsiella* spp., *Pasteurella* spp., *Pseudomonas aeruginosa*, *Streptococcus faecalis*, *Streptococcus uberis*, yeasts and molds (Blowey and Edmondson, 2010).

Most studies against bovine mastitis have focused on *Staphylococcus aureus*, considered a main pathogen for this disease. Because of the resistance to antibiotics observed in some strains, EOs and isolated components are used as alternatives to common antibiotics. Intriguingly, most investigations observed lower antimicrobial effect against the clinical isolates when using isolated compounds instead of EOs. This suggests synergistic interactions of the active components are important.

Some plants have already demonstrated their efficacy *in vivo*, such as *Ocimum sanctum*, which significantly reduced somatic cell count and ceruloplasmin concentration, decreasing udder inflammation connected with immunomodulatory effects in cows with subclinical mastitis.

Additionally, in an *in vivo* study, a strong antibacterial activity of EOs of *Thymus vulgaris* and *Lavandula angustifolia* against *Staphylococcus* sp. and *Streptococcus* sp. has been demonstrated. An intramammary application of these oils and of a mixture of them caused a drastic decrease in bacterial count in different samples of milk after 4 consecutive days of treatments. A stronger antibacterial activity was achieved by external application of these oils in Vaseline with a rate of recovery of 100% with thyme essential oil (Wells, 2024).

Also, low concentrations of EOs from *Coriandrum sativum*, *Origanum vulgare*, *Syzygium aromaticum*, and *Thymus vulgaris* reduced biofilm formation by at least 80% against clinical isolates from *Staphylococcus aureus* (Albuquerque et al., 2023).

EOs can also be used to control fungi, including the protothecosis *Prototheca zopfii*, which is a further common pathogen causing mastitis in animals (Lass-Flörl and Mayr, 2007). In a study, it was found that all tested strains of *P. zopfii* were sensitive to thyme, marjoram and oregano, which all contain carvacrol as an active compound.

Case in point: EOs and methane production from cows

A blend of EOs containing cloves, coriander seeds and geranium effectively modulated the ruminal microbiota towards more efficient fermentation, which reduced the total methane production of Holstein cows (Rossi *et al.*, 2022). In other studies, the addition of thyme EO to the basic ration reduced methane emissions, suggesting thyme mitigates methane production in ruminant cattle. Other EO blends containing thymol, eugenol, cinnamaldehyde and carvacrol in dairy feed decreased the population of protozoa, methanogens and proteolytic and biohydrogenase bacteria, which led to optimal ruminal fermentation and decreased methanogenesis (Daning *et al.*, 2020).

4.2.2 Bovine respiratory disease

Bovine respiratory disease (BRD) is a viral and bacterial disease affecting newly weaned or received cattle. *Mannheimia haemolytica, Pasteurella multocida, Histophilus somni* and *Mycoplasma bovis* are the main bacterial pathogens associated with BRD, which is considered to be the costliest health problem in the North American feedlot sector, as it is the leading cause of cattle morbidity and mortality, contributes to poor feedlot performance and carcass merit and is costly to treat (Duff and Galyean, 2007; Amat *et al.*, 2019). Investigations showed that thyme (carvacrol) and fennel displayed the strongest antimicrobial activity against these pathogens and have potential to mitigate BRD through intranasal administration.

Bacterial respiratory infections affecting pigs, such as pneumonia, pleuropneumonia and pleurisy, are a major health concern in the swine industry and are associated with important economic losses. EOs have the potential to be used as antimicrobial agents against major swine respiratory pathogens as alternatives to antibiotics (LeBel *et al.*, 2019).

Essential oils in general and thyme oil in particular can be used in veterinary medicine as a natural alternative to antibiotics, but oral administration should be limited to maintain the probiotic effect of these bacteria (Mancianti and Ebani, 2020). Regarding calves, EOs can have a positive impact pre- and post-weaning. Adding a blend of EOs (anise, cinnamon, garlic, rosemary and thyme) to milk replacer contributed to immunity improvement (as determined by weekly blood samples). This EO blend decreased morbidity of neonatal diarrhoea without affecting feed intake, animal performance characteristics, body development or blood metabolites (Palhares Campolina *et al.*, 2021).

Oregano oil fed to calves for the first 10 days effectively reduces the number, severity and duration of neonatal diarrhoea syndrome (Katsoulos *et al.*, 2017). A similar antibacterial effect is reported against bacterial lamb enteritis caused by *Escherichia coli*, followed by *Salmonella* species and then *Klebsiella* species (Darwish *et al.*, 2021).

Additionally, EOs can also be used as a dietary additive to improve growth performance and reduce the incidence of diarrhoea in weaned piglets (Canibe *et al.*, 2022; Hernández-García *et al.*, 2024).

Essential oils are used in poultry farming to improve animal health and meat quality. These include preparations from *Cinnamomum zeylanicum, Syzygium aromaticum, Cymbopogon citratus, Mentha × piperita, Ocimum basilicum* and *Pelargonium graveolens* because of their good antibacterial activity, when used alone or in combination, vs *E. coli* strains isolated from poultry with colibacillosis or infections with *Aspergillus fumigatus* (Ebani *et al.*, 2018).

The EOs from thyme, basil and oregano reduce the levels of gram-negative *E. coli* and *Salmonella* bacteria in the gut of broiler chickens (Khattak *et al.*, 2014). The EOs from cinnamon and clove showed remarkable antimicrobial activity against strains of *Salmonella enterica* serotypes Enteritidis and

Typhimurium isolated from poultry. Both could also be employed, alone or in combination, in a farm environment for disinfection (Ebani *et al.*, 2019).

There are also a few reports on the use of EOs in fish farming. The EO from *Pelargonium graveolens* could be used as a dietary supplement in aquaculture of carp and crayfish (Abdel Rahman *et al.*, 2020).

Studies or reviews describing therapeutic effects of EOs on other farm animals, such as sheep, goats or horses, are hardly to be found in the literature.

If one summarizes the literature in this field, there is strong evidence that EOs can be used as adjuvants to traditional antibiotic drugs such as tetracycline, doxycycline and tilmicosin to reduce the effective dose and decrease bacterial resistance. In particular, carvacrol and thymol as constituents of some EOs have demonstrated additive and synergistic effects when combined with each other or with doxycycline or tilmicosin against *P. multocida* and *M. haemolytica*, indicating these EOs can be used as a feed additive as a novel therapy against multi-drug resistant bacteria (Kissels *et al.*, 2017). Similar findings showed that carvacrol and thymol enhance tetracycline effects against *S. aureus* by inhibiting the activity of efflux pumps (Schmidt *et al.*, 2016).

This aspect deserves special attention, as a combination therapy of antibiotics and EOs could also win over farmers who are rather sceptical about phytotherapy, as antibiotic consumption decreases and the effects quickly become visible. Bovine spongiform encephalopathy (BSE) might be an example, in which antibiotics and the EO components may act synergistically, such as by affecting multiple targets, by physicochemical interactions and by inhibiting antibacterial-resistance mechanisms (Langeveld *et al.*, 2014).

The combination therapy with antibiotics and EOs also makes sense from an evolutionary point of view, as components of EOs inhibit the development of antibiotic resistance. It is known that additional stressors reduce the fitness of bacteria. This is caused by mutations in various genes that slow down evolution and inhibit growth (Hiltunen *et al.*, 2018).

Although there are numerous articles discussing the potential of natural compounds as alternatives to antimicrobials in animal production, only a limited number of studies provide robust clinical evidence. Furthermore, the synergism among different EOs could be exploited to maximize antibacterial activity and minimize concentrations needed to achieve the expected effect.

4.3   Legal Status of Essential Oils as Veterinary Medicinal Products or in Animal Feed

Essential oils are used in the treatment of animals as veterinary medicinal products, i.e. veterinary phytotherapeutics, usually in the form of traditional veterinary phytotherapeutics (European Union, 2016; Menoud *et al.*, 2024) or as ingredients or additives in supplementary feed or dietary supplements (Giannenas *et al.*, 2013; Zeng *et al.*, 2015).

Although partly comparable in application, the preparations with EOs in the form of veterinary medicinal products or animal feed differ significantly in regulatory terms. The possible applications also depend on the target animal species. Particularly in the case of food-producing animals, special requirements such as withdrawal periods for milk, meat or eggs may have to be observed with regard to the time of use, duration and documentation (Hallier *et al.*, 2013; Rychen *et al.*, 2017).

4.3.1   Essential oils in veterinary medicinal products

In Germany, before the EU Regulation (EU) 2019/6 on veterinary medicinal products came into force, the legal basis for veterinary phytotherapeutics was §39b of the Medicinal Products Act (Arzneimittelgesetz, AMG): Registration of traditional herbal medicinal products.

Since EU Regulation (EU) 2019/6 on veterinary medicinal products came into force, only the preamble contains a reference to, and Article 157 contains details on, the possibility of a Commission report on traditional herbal products used to treat animals to legalize veterinary phytotherapeutics in Europe again in a simplified authorization procedure.

Therefore, there is currently no way to register new veterinary phytotherapeutics in Europe. All currently available veterinary phytotherapeutics are registered or authorized based on previous legislation (Table 4.1).

4.3.2   Essential oils in animal feed: Classification as feed additives

The initial approval of feed additives in the EEC/EC/EU was made through Directive 70/524/EEC. This directive included a blanket approval of all known substances with aromatic properties. The only requirement: to be safe for animals, consumers, users and the environment.

In the context of various food and feed scandals (BSE, rotten meat, etc.), consumer protection has come into the focus of food law and all related areas since the beginning of the 2000s. With the entry into force of Reg. (EC) No. 1831/2003, notification had to be given for all feed additives on the EU market, including EOs, and these were then listed with precise designations for the first time.

Essential oils whose supporting companies withdraw from the project or whose evaluation by the European Food Safety Authority (EFSA) is negative, lose their provisional approval granted after notification and are no longer marketable within the EU feed sector.

Essential oils for which notification had not been given under Reg. (EC) No. 1831/2003 were never approved with their designation as feed additives and therefore initially had no formal status under feed law. A classification when used in animal nutrition would then result from the purpose of use.

Table 4.1. Herbal veterinary products containing essential oils. (Author's own table.)

Herbal veterinary medicinal product	Medicinally active ingredients	Animal species	Country[a]	Application method	Manufacture	Differentiation	Registration number
Amos zwarte zalf	camphora, sulphobituminosis ammonium (ammonium bituminosulfonate)	ruminants, horses, pigs, dogs, cats, rodents	NL	topical (skin)	Kommer-Biopharm B.V.	over the counter	REG NL 2304
Apiguard®	**thymol**	bees	ES	topical (beehive)	Vita Bee Health Limited	prescription only, pharmacy only	
ApiLife Var.®	**thymol, eucalyptus oil, campher, menthol**	bees (WT: 0 days honey)	DE	local	MEDISTAR Serumwerk Bernburg AG	over the counter	
Benacet aethericum acetate mixture	**campher, eucalyptus oil, rosemary oil,** arnica tincture, aluminium diacethydroxide (acetic alumina), aluminium potassium sulphate (alum)	horses, cattle, pigs (WT: 3 days edible tissue, milk)	DE	local (powder)	SaluVet GmbH	over the counter	

Continued

Herbal veterinary medicinal product	Medicinally active ingredients	Animal species	Country[a]	Application method	Manufacture	Differentiation	Registration number
Cai pan	peppermint oil	dairy cows	NL	topical (skin)	Hemrik Products	over the counter	REG NL 7230
Chevicet®-t 281 mg/ml	**eucalyptol, cineol**	pigeons (exempt from compulsory marketing authorization according to §60 AMG)	DE	supension	Chevita GmbH	over the counter	
Chevi-rhin	**eucalyptol, menthol, pure turpentine oil, essential thyme oil**	pigeons (exempt from compulsory marketing authorization according to §60 AMG)	DE	local (solution)	Chevita GmbH	over the counter	
ColoSan®	**Aetheroleum Anisi stellati, Aetheroleum Carvi, Aetheroleum Cinnamomi cassiae, Aetheroleum foeniculi amari,** sulfur	dogs, rabbits, horses, cattle, sheep, goats, pigs (WT: 0 days edible tissue, milk)	AT, DE, NL	oral (solution)	SaluVet GmbH	pharmacy only	

Product	Ingredients	Animals (withholding time)	Country	Route (form)	Manufacturer	Legal status	Registration
EucaComp®	**eucalyptus oil,** marjoram, lemon balm, calendula	horses, cattle, pigs (WT: 0 days edible tissue, milk)	AT, DE	intrauterine, vaginal (suspension)	SaluVet GmbH	prescription only, pharmacy only	
Eucalyptusöl N	**eucalyptus oil**	dogs, cats, horses, cattle, sheep, goats (WT: 0 days edible tissue, milk)	DE	oral, local, per inhalation	SaluVet GmbH	over the counter	
Euterbalsam Dr. Schaette	**camphor, eucalyptus oil, clove oil, rosemary oil,** St John's wort oil, **laurel leaf oil**	cattle, sheep, goats, horses (WT: 3 days edible tissue, milk)	DE	local (emulsion)	SaluVet GmbH	over the counter	
Feed farm kamferichthammmol zalf	camphora, sulphobituminosis ammonium (ammonium bituminosulfonate)	ruminants, horses, pigs, dogs, cats, rodents	NL	topical	Feed Farm B.V.	over the counter	REG NL 10097
GH 57-Salbe	**camphor, eucalyptus oil, turpentine oil**	horses, cattle, sheep, goats, zoo animals (WT: 3 days edible tissue, milk)	DE	local (ointment)	Alvetra LIVISTO	pharmacy only	

Continued

Table 4.1. Continued

Herbal veterinary medicinal product	Medicinally active ingredients	Animal species	Country[a]	Application method	Manufacture	Differentiation	Registration number
Kamfer-Ichtyolzalf	camphora, sulphobituminosis ammonium (ammonium bituminosulfonate)	non-lactating cattle	NL	topical (skin)	A.A.-Vet Diergeneesmiddelen N.V.	via the veterinarian	REG NL 4348
KlauSan® -Paste Dr. Schaette	**larch turpentine,** coneflower, mei	horses, cattle, sheep, goats, pigs (WT: 3 days edible tissue, milk)	DE	local (paste)	SaluVet GmbH	over the counter	
Melissengeestadem spray	lemon oil, clove oil, citronella oil, cinnamon oil, fennel oil, coriandrum oil, caraway seed oil, angelica oil, lemon balm oil	ruminants, horses, pigs	NL	topical, (nose)	EUROstyle B.V.	via the veterinarian	REG NL 5557
Original NJP liniment	cod liver oil, peppermint oil	dairy cows	NL	topical (skin)	Nardos A/S	over the counter	REG NL 7448

Product	Ingredients	Species	Country	Route	Manufacturer	Status	Registration
PhlogAsept®	witch hazel leaves, camomile blossoms, calendula blossoms, sage leaves, **thymol**	dogs, cats, horses, cattle, sheep, goats, pigs, poultry, rabbits (WT: 0 days edible tissue, milk)	DE	local (solution)	SaluVet GmbH	pharmacy only	
Sporyl	**Aetheroleum Caryophylli,** cetrimid	cattle (WT: 0 days edible tissue, milk)	AT	topical (skin)	Richter Pharma	prescription only, pharmacy only	
Thymovar®	**thymol**	bees	ES	topical (beehive)	Andermatt BioVet GmbH	prescription only, pharmacy only	
Uierbalsem	camphora, **eucalyptus oil, rosemary oil,** arnica tincture, clove oil, hypericum oil, bay leaf oil	ruminants and horses	NL	topical (skin)	EUROstyle B.V.	via the veterinarian	REG NL 5627

aNL = Netherlands, ES = Spain, DE = Germany, AT = Austria.

4.4   Inhalation of Essential Oils for Respiratory Diseases in Horses and Small Animals From the Practitioner's Point of View

4.4.1   Pulmonary application

Aerosols containing active ingredients used to treat respiratory diseases can be distributed directly at the site of application over a very large and well-perfused area, where they take effect immediately. Compared to the gastrointestinal tract and skin, the respiratory epithelium has a weaker barrier function. The first-pass effect is bypassed, which results in better bioavailability than with oral application and enables a dose reduction. This in turn improves tolerability, reduces systemic side effects and increases drug safety.

4.4.2   Inhalation of essential oils

These are all advantages that make the inhalation of EOs particularly interesting as a therapeutic measure. Due to their broad spectrum of action, which includes increased blood circulation, increased secretion of seromucosal glands, mucolysis, bronchospasmolysis, activation of the ciliated epithelium and antimicrobial activity, EOs are predestined for the treatment of respiratory diseases. Nevertheless, they are not currently considered in veterinary medicine. The traditional application is being forgotten. Practical inhalation methods for EOs that have been modernized according to recent scientific findings do not yet exist.

4.4.3   The history of inhalation

The possibility of curing respiratory diseases by inhaling vapours and smoke from medicinal plants has fascinated people for thousands of years. Information on the medical use of inhalation can be found in numerous cultures (Sanders, 2007; Mehta *et al.*, 2018). Inhalation in animals is mentioned in the first veterinary writings, the hippiatric works of late antiquity (Von den Driesch and Peters, 2003). Inhalation is also described in the literature of the 'Stallmeister period' (1250–1762). In the *Nachrichters Roß-Arzneybuch*, first published in 1716, Deigendesch recommends smoking wormwood herb in a pan heated on a fire to treat strangles (druse) in horses, a highly contagious *Streptococcus* infection of the upper respiratory tract (Deigendesch, 1785). In 1866, Röll (1818–1907), the founder of Austrian veterinary medicine, described both steam inhalation and the smoke that develops when tar, resins and gum-resins are heated or burnt and placed on a red-hot iron or poured into water into which a red-hot iron is pushed (Röll, 1866) as common forms of application for medicinal plants containing EOs in one of the first pharmacopoeias for veterinarians.

Almost 30 years later, in his *Arzneiverordnungslehre für Tierärzte*, Fröhner (1858–1940) dealt in detail with the inhalation of gaseous, vaporous and powdered drugs and distinguished between direct action on the respiratory organs and general action after prior absorption. Fröhner had already reported on experiments that prove that fine solid bodies and all drugs that change into gaseous form, especially ether, easily reach the alveoli, while water vapours are deposited in the upper respiratory tract. At this time, inhalation boxes modified from human medicine were already being used for smaller animals in the clinics of the Berlin Veterinary College. Large animals inhaled water vapour using tube-like masks. Among other applications, hay vapours were used to loosen mucus in the bronchi (Fröhner, 1896). If, in addition to the expectorant effect stimulating the ciliated epithelium, an antiseptic effect is also to be achieved, Fröhner recommends inhalation with turpentine oil in a 1–5% mixture with water or inhalation with peppermint oil, which he also mentions as a therapeutic trial for pulmonary tuberculosis (Fröhner, 1896).

With the advent of inhalation anaesthesia in animals in the second half of the 19th century, the development of masks suitable for inhalation also began for various animal species (Von den Driesch and Peters, 2003). However, there was no further development of inhalation for the treatment of respiratory diseases.

4.4.4 Proven formulations

Both the manuals designed by veterinarians for animal owners and the textbooks for veterinarians from the 19th century to the 1960s contain tried and tested formulations for animals inhaling EO-containing drugs and EO mixtures. As late as 1968, a collection of veterinary prescriptions for pharmacists was published with proven formulations for inhalation to treat respiratory diseases of various animal species (Table 4.2; Stather and Döderlein, 1968).

Table 4.2. Formulations for inhalation to treat respiratory diseases of various animal species. (Author's own table.)

Bronchial catarrh in horses	Laryngitis, pharyngitis (angina) of the dog	Respiratory diseases of poultry
Recipe Oleum Pini pumilionis Oleum Eucalypti aa 1.5 Oleum Therebinthinae 30.0 to vaporize with boiling water.	Recipe Menthol 0.5 Oleum Pini pumilionis Oleum Eucalypti aa 1.5 Oleum Therebinthinae 30.0 for inhalation; a few drops twice daily in hot water.	Recipe Menthol 0.5 Oleum Eucalypti Oleum Therebinthinae aa 10.0 for inhalation; 10 drops in 1 l of hot water.

Some of these, as well as older formulations, have been taken up in the literature of the last 15 years on the practical use of medicinal plants in animals, designed for animal owners and veterinarians, after they have been subjected to phytopharmacological assessment according to new toxicological findings (Aichberger *et al.*, 2012; Reichling, 2016; Brendieck-Worm *et al.*, 2021; Brendieck-Worm and Melzig, 2021).

In particular, inhalation with chamomile flowers and chamomile tincture, eucalyptus oil, spruce and pine needle oil, peppermint oil, herb and oil of thyme (*Thymus vulgaris*) and the less eye-irritating EO of *Thymus* ct. linalool are recommended here.

4.4.5   Steam inhalation of EO in folk medicine and traditional medicine

Steam inhalation is still used sporadically in folk medicine and empirical medicine. This is confirmed by surveys of farmers in the Mediterranean region, Switzerland, Austria and Bavaria conducted as part of ethnoveterinary research (Schlittenlacher, 2022).

4.4.6   Steam inhalation of EOs in veterinary practice

In the veterinary practice of the author Ms Brendieck-Worm, steam inhalation has been used successfully, particularly for respiratory diseases in wild animals including wild birds, young animals and uncooperative pets, especially cats. Vapour inhalation has also proved successful in animals in a particularly critical state of health, as it is associated with considerably less stress than forced injection treatment or oral administration of medication. The prerequisite for this is the willingness of the animal owner to carry out such inhalation 2–3 times a day, or more often if necessary, following appropriate instructions.

The animals are restrained in well-ventilated cages and protected from contact with the hot inhalant. An inhalation room as small as possible is created for the steaming inhalant and the cage. An infusion with chamomile is generally used.

4.4.7   The end of phytotherapy in veterinary medicine in the 20th century

Despite positive therapeutic experience and proven effects, in veterinary medicine in the 20th century, labour- and personnel-intensive inhalation, particularly with EOs for the treatment of respiratory diseases, took a back seat to injection treatment and the use of orals, and disappeared from the textbooks with the advent of antibiotics and steroidal and non-steroidal anti-inflammatory drugs, apart from a few anecdotal mentions without specific instructions for use. In the actual standard textbook on veterinary pharmacotherapy used at all German-speaking veterinary universities,

---

Case in point: Siamese tomcat Jimmy, age: 5 months

Case history

The male cat has been undergoing veterinary treatment for catarrhal rhinitis with increasing inspiratory and finally expiratory rales since the age of 4 months without success. He was referred to a veterinary clinic on suspicion of a growth in the nasopharyngeal region.

Diagnosis

Retropharyngeal polyp the size of a cherry stone.

Therapy

Extensive surgical removal, electrolyte infusion, dexamethasone, enrofloxacin, meloxicam, tramadol.

Course

Discharge from the clinic after 24 hours. The following day, the cat appears seriously ill for follow-up treatment at the surgery. He is hospitalized again in the clinic. The owner picks him up again after 4 days in the same condition and brings him back to the practice immediately.

The cat is powerless and weak-willed, suffers from severe respiratory distress, eats and drinks nothing.

Therapy

Discontinuation of all previous medication, outpatient 2 × daily s.c. fluid and energy intake, organotherapy (Membrana nasalis comp. PLV, Larynx/Apis comp. PLV, Bronchi comp. PLV) s.c., at home hourly chamomile steam inhalation.

Course

Breathing gradually normalizes after 3 days. The cat begins to groom himself and purr, has some appetite. On the 5th day, food and water intake are normal and the general condition improves daily.

In the following months, slight respiratory problems occur again and again, which are treated with chamomile steam inhalation. As the head skeleton grows out, the symptoms largely disappear.

---

phytotherapeutics, including EOs, are presented as insufficiently scientifically proven in terms of efficacy and therapeutic safety and their use is discouraged (Löscher *et al.*, 2014). Unfortunately, there is no interest in using their potential for the urgently needed reduction of antibiotics and for the containment of increasingly occurring chronic respiratory diseases.

4.4.8 Inhalation in current veterinary medicine

Inhalation as a form of application has only come back into focus in veterinary medicine in the past 20 years with the increasing incidence of chronic respiratory diseases. In horses this is chronic obstructive bronchitis (COB), in dogs, chronic bronchitis (CB) and chronic eosinophilic bronchopneumopathy (EBP) and in cats, feline atopic syndrome (FAS). These diseases are currently mainly treated with glucocorticoids and synthetic bronchodilators. In view of the serious side effects of systemic glucocorticoid administration, inhalation is gaining in importance. The experience gained in human medicine with this form of administration is being used here. In dogs, cats and horses, spacers are currently used in combination with metered-dose inhalers. This form of inhalation requires long-term cooperation from the animal. Good advice and guidance for the pet owner administering the inhalation is just as important here as the patient and low-stress familiarization of the animal to be treated with the inhalation system. Good instructions for inhalation in cats can be found at the website of International Cat Care (n.d.).

The efficacy and relative safety of inhalation treatment of horses suffering from COB with corticosteroids and bronchodilators has now been proven in numerous studies (Niedermaier and Gehlen, 2009). The good efficacy of inhaled glucocorticoid administration has also been proven for cats with FAS and dogs with CB and EBP and has been established as a treatment option, although there are currently neither approved glucocorticoids for inhalation nor tested dosage recommendations for veterinary medicine, and treating veterinarians are dependent on personal experience (Klenk and Schulz, 2022).

4.4.9 Inhalation of essential oils as an antibiotic-reducing therapy?

Inhalation with EOs for the treatment of infectious respiratory diseases is still waiting for a contemporary comeback. There is still an urgent need for action in the development of antibiotic-minimizing therapy methods. The potential of EOs as anti-infectives and at the same time to support the self-healing powers of the organism is clearly demonstrated by the studies on EO drugs, such as oregano, initiated by the animal feed industry following the ban on antibiotic performance enhancers in 2006.

4.4.10 Hygienizing and improving indoor air with EOs

The coronavirus pandemic in 2020 is not the first time this has been known: ventilation and air exchange have a decisive influence on the spread of viruses and therefore on the risk of infection. Suitable ventilation technology can help to reduce infection through aerosol-bound transmission of viruses, bacteria and fungi. Using evaporated EOs enables indoor air to be kept clean and ventilation systems to be sanitized. In comparison to the substances currently used for room disinfection, EOs are generally less toxic (Pibiri *et al.*, 2006; Banovac,

2012). Mixtures of EO with antiseptic and disinfectant properties to reduce the germ content in the air of stables have been on the market for a long time and include eucalyptus oil, tea tree oil, thyme oil and citronella oil. They can reduce the microbial load in the indoor air and thus contribute to respiratory health.

4.4.11 Determination of antimicrobial efficacy

EOs are highly potent natural substances and stimulate the mucous membranes even at low doses. For therapeutic use in infectious diseases of the respiratory tract, it is therefore essential to know both the general sensitivity of the pathogens to the EO used and the lowest effective dose and its tolerability in the animal.

An aromatogram provides an initial overview of the sensitivity of the pathogens involved in the pathological process in the respiratory tract (Müller and Schneider, 2015). The aromatogram has been established in veterinary medicine for several years. However, the range of EO tested for veterinary pathogens is still relatively small (Bismarck *et al.*, 2017). In addition, pathogens of respiratory tract infections are still poorly considered. If the therapeutic dosage of an EO tested as effective in the aromatogram is to be determined, a microdilution test can be used, but this is much more costly and time-consuming.

4.4.12 Species differences – Problems in practical application

4.4.12.1 Respiratory rate and tidal volume

When administering inhalation therapy to animals, considerable functional and morphological differences between animal species must be considered, which affect the particle deposition of the inhalate in the airways to varying degrees.

Different breathing patterns of various animals include the following:

- Small pets and birds have high-frequency breathing (e.g. rabbits, 30–60 breaths/min, guinea pigs 40–100 breaths/min) with a small tidal volume.
- Horses have a frequency of 8–14 breaths/min and a tidal volume of up to 180 l/min.

In some cases, considerably higher respiratory rates are measured in young animals than in adults.

The forced breathing with high flow velocities in small animals favours the precipitation of aerosol in the upper respiratory tract and thus makes it more difficult for the particles to penetrate into the lower sections of the lungs (Schulz *et al.*, 2003).

4.4.12.2 Nasal breathing

The dose of aerosol reaching the alveoli is significantly lower when breathing through the nose than when breathing through the mouth, as demonstrated by studies in humans. Particles > 3 µm are already effectively filtered out in the upper respiratory tract during nasal breathing. This is particularly important in horses, which are only capable of nasal breathing (Reinhold and Fehrenbach, 2003).

4.4.12.3 Inhalation systems

Nebulizers (jet, ultrasonic and membrane nebulizers), pressurized gas metered-dose inhalers, normal pressure metered-dose inhalers and powder inhalers are currently available for inhalation. Inhalers in which steam is generated with hot water and volatile pharmacologically active substances are also available. All inhalation systems produce aerosols with particles of different sizes, with steam nebulizers producing relatively larger particles than cold nebulizers.

Steam inhalation is therefore reserved for problems of the upper respiratory tract. It is currently the only accepted form of application of oil inhalation but is considered more of a household remedy without great therapeutic value. Manufacturers of jet and ultrasonic nebulizers generally exclude the use of EOs in their devices at present. There are also numerous rumours circulating on social media about the harmfulness of EO inhalation.

4.4.13   Requirements for establishing EO inhalation as a valuable therapeutic option

In order to be able to use the therapeutic potential of EO inhalation effectively in the future, especially for respiratory tract infections, efforts are needed from various sides:

- On the part of university pharmacology, the negative attitude towards the use of natural complex substances, in particular EOs, should be abandoned and the utilization of these substances accepted as a research task.
- The dose-dependent tolerance of EOs on the epithelia of the respiratory tract in different animal species should be determined – also considering an increasing hyperreactivity.
- The antimicrobial efficacy of at least the EOs traditionally used for respiratory diseases should be investigated and the determination of their minimum inhibitory concentration in veterinary-relevant pathogens of respiratory diseases should be determined.
- User- *and* animal-friendly inhalation systems should be developed for the needs of the various species.

EOs support and activate the body's defence mechanisms and self-healing powers. Used preventively and in the early stages, their antimicrobial effect in combination with immune stimulation/modulation, bronchospasmolysis, mucolysis and stimulation of secretomotor activity can make a significant contribution to avoiding the use of antibiotics in respiratory diseases and preventing chronic lung changes.

Aromatherapy: How Essential Oils Can Help with Modern Human 'Epidemics' – From the Practitioner's Point of View

5

Eliane Zimmermann*

5.1 Insights Into 35 Years of Experience in Clinical Aromatherapy

As genuine essential oils (EOs) are by definition mixtures of multiple substances extracted from a plant, they can offer a wide range of highly interesting therapeutic effects. Studies have shown that they and also a number of their isolated constituents have versatile effects. I can confirm the following findings from my 35 years of clinical aromatherapy:

- They are effective against a large number of pathogens, with a broad-spectrum effect: thyme chemotype (ct.). thymol, tea tree, oregano, clove, cinnamon (Tanasă *et al.*, 2024).
- They interfere with the quorum sensing of pathogenic germs: rose geranium, Damask rose, ginger, marjoram, lemongrass (Camele *et al.*, 2019; Maggio *et al.*, 2025).
- They have anti-inflammatory activities: German chamomile, yarrow, copaiva, rose absolute.
- They have immunomodulatory effects, stabilizing mast cells: Atlas cedar and Himalayan cedar (Gupta *et al.*, 2011; Chauiyakh *et al.*, 2023).
- They can modulate the release of cortisol and oxytocin, i.e. support symptoms of anxiety and stress: clary sage, orange, rose geranium (Hans *et al.*, 2023; Nakajima *et al.*, 2024; Nascimento *et al.*, 2024).
- They can be slightly antihypertensive: ylang-ylang and sandalwood (Jung *et al.*, 2013; de Groot and Schmidt, 2017).

*Corresponding author: zimmermann@aromapraxis.de

- They can modulate gamma-aminobutyric acid (GABA) receptors: lavender (Wang and Heinbockel, 2018).
- They can support secretolytics, especially for ENT (ear, nose and throat) complaints: myrtle, eucalyptus (*E. radiata*), ravintsara, cardamom (Ciuman, 2012).
- They can have spasmolytic effects, especially on the smooth muscles: bishop's weed, sweet marjoram (Heghes *et al.*, 2019; Shahrajabian *et al.*, 2021).
- They can be anticarcinogenic, antitumoral (mainly *in vitro* studies): orange, bergamot, tea tree (Bhalla *et al.*, 2013).

5.2   Case Report: Amputation Avoided

A woman who is familiar with essential oils, yet has no in-depth experience in dealing with serious illnesses, asked my colleague and me for advice. She was urgently seeking help because her brother (67) was due to have a leg amputated. After an accident, he had a severe flesh wound that became badly infected; it showed no significant reaction to the medication administered in the hospital for many weeks. As we already knew of two cases of colleagues who were able to prevent impending amputations through the customized use of a few essential oils, we encouraged her to try something similar. Her brother was more than grateful for all the advice, willing to follow a strict protocol with essential oils at home, so he left the hospital at his own risk.

We, the team of colleagues, considered the possibility of the wound worsening to be extremely low. Judging by former, similar experiences, rapid healing – at least partly – was rather to be expected. So, the dedicated sister started to clean the wound carefully before applying a wound dressing several times a day, using different mixes and concentrations of EOs. Her brother's personal olfactory preferences were considered; we encouraged him to choose mainly from EOs that were acceptable to his nose. He also took custom-made capsules with essential oils orally.

They alternated his topical treatment with two different mixes:

spray (shake well before use!)
Rosa × damascena Herrm. hydrosol 100 ml with 5 drops of each:
Leptospermum scoparium J.R.Forst. & G.Forst. (manuka) essential oil
Melaleuca leucadendra L. (cajuput)
Coriandrum sativum L. (coriandrum seed oil)
Cymbopogon flexuosus (Nees ex Steud.) W.Watson (lemongrass)
Thymus vulgaris ct. linalool L. (thyme ct. linalool)
Lavandula angustifolia Mill. (lavender)
Citrus limon (L.) Osbeck (lemon)

oil mix

> 50 ml St John's Wort (*Hypericum* macerated in olive oil) with 10
> drops of each:
> *Cymbopogon martini* (Roxb.; palmarosa)
> *Melaleuca alternifolia* (Maiden & Betche) Cheel (tea tree)
> *Thymus vulgaris* ct. thymol L. (thyme ct. thymol)
> *Leptospermum scoparium* J.R.Forst. & G.Forst. (manuka)
> *Citrus × aurantium* L. flos (neroli)

We chose the essential oils according to our experience, and according to evidence from countless aromatograms by German pharmacist Dorothea Hamm (Karlsruhe), reported in her book *Aromapraxis Heute* (Beier *et al.*, 2022), and personally. *Hypericum perforatum* L. was chosen for its wound-healing properties; the antibacterial and anti-inflammatory properties are important in the treatment of chronic wounds (Schempp *et al.*, 2002; Schempp, 2011). *Cymbopogon martini* Roxb.: Due to the main constituent geraniol (82%), palmarosa is a mild antibacterial essential oil which can inhibit many different pathogenic agents; it also enhances the anti-inflammatory action of *Hypericum* (Lin *et al.*, 2021). *Melaleuca alternifolia* (Maiden & Betche) Cheel: This essential oil is one of the best examples of complex natural mixtures; its combination of about 35% terpineol-4 along with 30% of various monoterpenes can almost promise a quick recovery from infectious sites. There is a lot of evidence for its broadband-like action against most common pathogens, from bacteria to fungi and viruses (Carson *et al.*, 2006; Battisti *et al.*, 2024; Dontje *et al.*, 2024). *Thymus vulgaris* ct. thymol L.: Adding the essential oil of this ancient European medicinal plant proved to be extremely helpful for avoiding the forming of dangerous biofilms and thus bacterial resistance, so helping the affected person's immune system to successfully heal its wounds. This essential oil contains around 35% thymol and carvacrol in varying proportions; both are well-known antiseptics (Mohammed Aggad *et al.*, 2025). *Leptospermum scoparium* J.R.Forst. & G.Forst.: Like its 'counterpart' from Australia, this gentle tea tree oil from New Zealand proved to help with various pathogens, most of the time with very convincing results. Its unique action is attributed to rare triketones like leptospermone and flavesone (up to 25%); they are valued for their antimicrobial and anti-inflammatory properties (Pedonese *et al.*, 2022) *Citrus × aurantium* L. flos: We like to add this oil for patients with bad wounds combined with heavy pain, therefore anxiety, and a general feeling of discomfort. It showed a good effect against multiresistant pathogens, which we had to avoid at all costs. Furthermore, it contributes to a more pleasant smell of our mix. It contains more than 50% linalool, some 20% monoterpenes and about 6% monoterpene esters like neryl acetate and linalyl acetate. They help in balancing cortisol, contributing to an uplifting effect (Choi *et al.*, 2014; Khodabakhsh *et al.*, 2015; Beier *et al.*, 2022).

A clearly recognizable improvement at the wound edges started as early as the next day. A short time later he was discharged from hospital, and after about 3 weeks the wound had shrunk from 12 cm to 5 cm. A month later it was at 1.5 cm. The man still has his leg, despite his consultant still opting for an amputation at this point, as there is still a biofilm. His blood work regarding inflammatory values is nearly normal at the time of writing.

5.3   Introducing Khella Essential Oil

Let me introduce a relatively unknown essential oil for the following cases. Khella, also called bishop's weed (*Ammi visnaga* L.; Fig. 5.1), is a close relative of the equally unfamiliar wild carrot (*Daucus carota* L.), which is even considered an undesirable weed in middle Europe (Khalil *et al.*, 2020). Khella needs warmth, even heat, and therefore grows in warm countries. Provided with these conditions, it is a vigorous plant that thrives in sandy, well-drained soils.

Fig. 5.1. Cultivation of khella, also called bishop's weed (*Ammi visnaga* L.), with blossoms and seed heads. (Author's own image, courtesy of WALA Heilmittel GmbH.)

Most khella essential oil comes from Morocco, which may be a reason why it is more popular with French aromatherapists (El Hachlafi *et al.*, 2023).

It looks very much like the pretty wild carrot, which is more well-known by the beautiful name 'Queen Anne's Lace'. Both display attractive white umbel-like inflorescences which are composed of dozens of rays, each bearing tiny umbels with white flowers. Those rays can turn quite rigid and woody, so they come in handy when a toothpick is needed, hence another common name: toothpick ammi. Both plants provide us with tiny aromatic seeds which can be distilled.

Khella essential oil is usually more expensive. Its fragrance reminds me a bit of parsley, with earthy and slightly sweetish undertones. Some people describe it as aniseed-like. Its very pronounced mode of action is due to a high content of butyrates:

- almost 20% isoamyl-2-methylbutyrate;
- around 10% each of amylisobutyrate and amylvalerate; and
- small amounts of isoamylisobutyrate, iso-butyl-2-methylbutyrate, 2-methylbutyl-2-methylbutyrate, 2-methylbutylisobutyrate and isoamylvalerate.

These compounds add up to around 50% of the total. They are extremely helpful when the smooth muscles of our inner organs are tense. Some 30–38% linalool plus small amounts of α-thujene, α-pinene, β-pinene, β-myrcene and pulegone might contribute to the surprisingly effective topical applications of this oil (Kamal *et al.*, 2022). As we cannot consciously control our smooth muscles – they're controlled by the nervous system involuntarily – it may be encouraging to find such medications in case of painful spasms, especially of the bronchi and the intestines, but also of heart muscle tissue.

'Tightness of the chest', a condition which is often heard of after COVID infections, responds well to a treatment with khella essential oil (of course the individual conditions have to be checked by medical professionals beforehand). Patients experience a deep alleviation when it is applied around the chest (diluted at around 5–10% in a fixed oil), but also used for inhalation, for example in an aromastick. Those applications lead to quick relief for patients with anxiety and the impression that there is not enough oxygen reaching their lungs.

More than 15 years ago, I had sniffed khella essential oil for the first time and learnt from a dear colleague that she had managed to relieve bad asthmatic fits of her little son with this oil. It was completely unknown to me at this point. Some 10 years later, (indirect) scientific findings were published by Professor Hanns Hatt and his team from the University of Bochum. They provided the first evidence to show that olfactory receptors are functionally expressed in human airway smooth muscle cells and regulate pathophysiological processes (Kalbe *et al.*, 2016). They stated in 2016: 'Therefore, olfactory receptors might be new therapeutic targets for these diseases, and blocking olfactory receptors could be an auspicious strategy for the treatment of early-stage chronic inflammatory lung diseases.'

This could be good news for asthmatics. Two types of olfactory receptors can be found in the muscle cells of the human bronchi. Those receptors on the muscle cells were labelled OR2AG1 and OR1D2. The researchers also identified the scents that match the olfactory receptors and the signalling pathways that trigger them in the cell (Kalbe *et al.*, 2016).

Amyl butyrate, with a fruity odour with notes of banana and apricot, activates the OR2AG1 receptor. When the odorant binds, it relaxes and dilates the bronchi. The effect in their experiments was so strong that it was able to neutralize the effect of histamine. The body releases this substance in allergic asthma, causing the bronchial tubes to constrict. The researchers also showed that amyl butyrate triggers the same signalling pathways in the muscle cells as in the olfactory cells of the nose. 'Amyl butyrate could help improve airflow in asthma', concludes Hanns Hatt. 'It can probably not only counteract the effects of histamine, but also those of other allergens that affect deep breathing' (Kalbe *et al.*, 2016).

The receptor could also be of interest for the treatment of other diseases, such as chronic obstructive pulmonary disease (COPD). If the right odours activate these receptors, the bronchial tubes dilate or constrict – a potential approach for asthma therapy and further lung diseases which cause breathlessness, according to Hanns Hatt.

But after hearing those findings, I first checked the reference book that I wrote almost 30 years ago (first published in 1998, now in its revised 7th edition). It contains a comprehensive list of the main components of almost 200 essential oils (Zimmermann, 2022). I found that khella essential oil holds the record for containing butyrates, which otherwise are more minoritarian and random molecules in essential oils. My little list showed the following:

- Khella (*Ammi visnaga*) can contain about 50% different butyrates, i.e. esters of butyric acid.
- Roman chamomile (*Chamaemelum nobile*), which is known for its strong antispasmodic effect, contains around 13% of these compounds derived from butyric acid.
- The three chemotypes of myrtle essential oil (*Myrtus communis*: 1.8-cineole, myrtenyl acetate and α-pinene), helichrysum (immortelle, *Helichrysum italicum*), lavender (*Lavandula angustifolia*) and palmarosa (*Cymbopogon martini*) each contain trace amounts of butyrates. Yet experience by many practitioners has shown that small amounts of these compounds seem to be sufficient for an essential oil to have a pronounced relaxing effect.

Butyrate – one of several short-chain fatty acids – is produced by 'good' bacteria in your gut and is the main source of energy for intestinal cells so that they can break down dietary fibre in the large intestine (colon). It helps in supporting our immune system, reducing inflammation and preventing diseases like cancer (Hamer *et al.*, 2008; Liu *et al.*, 2018).

As amyl butyrate, also known as pentyl butanoate, is a molecule similar to some of the constituents in khella essential oil, my curiosity as a clinical aromatherapist was aroused. Therefore, I transferred those new findings to people familiar to me, recommended the use of khella essential oil for complaints after COVID infections and got impressive feedback.

5.4   Case Reports: Back to Better Breathing

5.4.1   Case 1

Soon after investigating the theory behind this quite unknown plant, I got an early telephone call from a neighbour who was wheezing and trying to reach help, as his breathing was very difficult, and he also complained about an unusual pain in his chest. I proposed calling the ambulance immediately, but this rather stubborn person didn't want to.

I grabbed my little bottle of khella essential oil along with some lavender, roman chamomile and eucalyptus, and rushed to his home. The situation didn't look good. I felt uneasy and tried to convince him again to go to the hospital. No way. So I rubbed some 5 drops of khella essential oil – just diluted in some organic olive oil in the palms of my hands – around his chest. And I handed him my bottle with the instruction to try to inhale the oil as slowly and as deeply as possible. I reminded him that if there was no improvement in about half an hour, I would call the ambulance, no matter what he said.

I sat near him and watched him as he slowly started to relax and to breathe more normally. After some 15 min he asked me: 'Is it possible that this stuff helps me breathe better again?' I have to admit that I was quite surprised, and of course relieved too.

He continued to inhale the oil for about a further 15 min and looked much better. After a long nap, he was almost back to his old self.

Being a podcaster, I described this 'miracle' in one of the episodes, which I record and publish together with a colleague. Thereafter, people with heavy breathing difficulties following an infection with COVID-19 came to my colleague's premises asking for help. Most of them absolutely fell in love with the fragrance of khella essential oil (whereas most healthy people are not too fond of its smell).

5.4.2   Case 2

Once, a wheezing and coughing nurse, our colleague who had persisting symptoms of Long COVID, came for help. She was reluctant to ask her GP for yet another course of hydrocortisone, as she has to take strong medications for another health condition. Of course, khella was the first aid recommendation, and she absolutely loved the oil. After just a few topical applications and inhalations, her breathing was getting much better and was almost normal.

We recommend mixing it with some other oils, which also have a relaxing effect:

- 10 ml *Simmondsia californica* (Link) C.K.Schneid. (jojoba oil);
- 2 drops *Ammi visnaga* L. (khella);
- 2 drops *Citrus sinensis* (L.) Osbeck or *Citrus reticulata* Blanco (sweet orange or mandarine);
- 1 drop *Chamaemelum nobile* (L.) All. (Roman chamomile); and
- 3–4 drops *Vanilla planifolia* Jacks. ex Andrews (ethanolic extract of organic vanilla fruits).

For inhalation purposes, we recommend 3 drops of khella in an aromastick; 1–2 drops of Roman chamomile and 3–4 drops of mandarine essential oil might be added so the fragrance gets really pleasant.

Recently we got a very positive feedback from a former student of ours. She is also a nurse, who recommended it to two other nurses – both very sceptical; they didn't believe in natural help for the bad wheezing that had affected them after COVID infection (both non-smokers around 60 years old). After they had used the aromastick with the khella mix just for a few days, one of them approached our former student enthusiastically, almost embracing her, as she absolutely loved the smell. Her breathing had improved substantially, as had the breathing of the other colleague.

Another convincing case report was that of a grandmother and her 4-year-old grandson who had suffered from recurrent acute spasmodic lar-yngitis since he was 6 months old. The attacks occurred at least once a week and he was being treated with hydrocortisone. As khella essential oil seems to have favourable properties for bronchitis and asthma, she and her daughter decided to apply an ointment with khella twice a day to the affected area, i.e. the anterior neck, with promising results.

To summarize, khella essential oil has a strongly relaxant/spasm-relieving effect on the smooth (involuntary) muscles; it opens and relaxes in spastic bronchitis with a vasodilating effect. In France it is recommended as an adjuvant for coronary heart disease. Furthermore, khella has also been described as helpful for liver and kidney colic, psoriasis (with carrot seed oil) and vitiligo (Fu *et al.*, 2020; Abukhalil *et al.*, 2021). In these times, with many COVID infections still occurring, plus all of those 'old-fashioned' viral chest infections, with the almost inevitable 'chest tightness' in many patients, I thought the 'aromatic world' should know that there is a con-vincing recourse for those who wish to help their bodies heal with natural substances. Khella essential oil seems to be a 'magic wand' when those symptoms are prevalent.

5.5   Case Report: The Threat of Cervical Cancer Averted

When you receive an email starting with 'thank-you-thank-you-thank-you', curiosity gets sparked: Once again I got the confirmation that essential oils really do work in cases of scary Pap test results. Despite the vaccination programmes in place, the cervixes of too many young women are affected by lesions which can lead to cancer, caused by infections with the human papillomavirus (HPV). In those cases, I dearly wish that many more women knew about the specific antiviral essential oils I usually recommend for those patients! This woman, who was one of several women seeking help, and who was very frightened, wrote to me: 'We wrote in February regarding the vaginal suppository prescription and the HPV virus. Mid-October, a letter from the Women's Hospital arrived. What can I say ... I am HPV negative. And yes, I decided against the vaccine and in favour of the vaginal suppositories according to your prescription.'

We get similar emails or letters regularly. In one of our podcast episodes, we had interviewed a young woman who wanted to give her body a chance to heal using essential oils based on suggestions in our book, even though her gynaecologist was urging her to use a different treatment method. We talked about her experiences and about her healing process. Since then, we have had regular feedback, many of the affected women reporting that there can be a significant improvement after 6–8 weeks, which is easily verifiable by the gynaecologists. More than ever, we believe that women should give themselves this chance, as the HPV health issue is obviously extremely common – in times of compromised immune systems and autoimmune attacks, probably even more so now than a few years ago. The diagnosis terrifies countless women, especially when the (supposedly) 'bad news' is neither delivered at eye level nor with empathy, and usually without any recommendations to try a natural approach first. Incidentally, such an infection goes unnoticed by many women, and so does the healing – thanks to the amazing efforts of the human immune system. However, there are of course exceptions.

There are about 40 genital HPV, which can not only cause cervical cancer, but also other HPV-related cancers such as penile, anal, vulvar, mouth and throat cancer. It is the task of a functioning immune system to keep these germs, like others, in check on a daily basis. This is the case with an individually balanced microbial flora in which the non-pathogenic germs dominate. Experience has shown that the natural fragrances mentioned can help with this.

Affected women should definitely encourage their sexual partners to use the antiviral oil mixture too: twice a day (for around 6 weeks) once they have been diagnosed or warned by their gynaecologist. Otherwise, there will be an 'eternal' ping-pong of pathological germs. I recommend a mix of 2 drops each of 3 of the following essential oils in 10 ml of fixed oil to be applied gently to the genital area drop by drop. Women should ideally apply a few drops on the tip of a small tampon too and leave it overnight. This procedure should be done for approximately 4 weeks followed by a pause of 2 weeks. Then the daily external application only can be done for as long as 6 months. The mix should be slightly different after 2–3 weeks.

In 10 ml of virgin olive oil:

- *Rosa × damascena* Herrm. (rose);
- *Melissa officinalis* L. (lemon balm);
- *Cistus ladanifer* L. (cistus/rockrose);
- *Leptospermum scoparium* J.R.Forst. & G.Forst. (manuka);
- *Melaleuca alternifolia* (Maiden & Betche) Cheel (tea tree); and
- *Salvia rosmarinus* ct. verbenone Spenn. (rosemary ct. verbenone).

5.6 Future Breeding Grounds for Modern Epidemics – Microplastic as One Example

Increasing exposure to environmental hazards such as endocrine disruptors (EDCs), non-intentionally added substances (NIAS) and microplastic presents new challenges for our immune system (Peters *et al.*, 2019; Nerín *et al.*, 2022; Ghosh *et al.*, 2023; Stiefel and Stintzing, 2023). Furthermore, the environment, with its complex food chains, is not unaffected. In particular, as the contamination of the human body with micro- and nanoplastics and the resulting attraction of microorganisms is increasing (Li and Liu, 2024), the only effective treatment for such infections in the near future might happen with natural complex substances like essential oils – due to the above-mentioned properties.

As published at a 3-day international conference on this topic at the University of Dublin in late summer 2024, science now has the imaging techniques at its fingertips to prove the devastating consequences of the increasing daily littering with plastics of our environment, of the human body, and of animals too. Every single plastic bottle (mostly PET; polyethylene terephthalate) with 'drinking water' and soft drink can contaminate us with up to 240,000 particles of plastic (Dolcini *et al.*, 2024). Each of these particles in turn acts like a magnet for pathogenic germs and can therefore lead to untreatable infections.

The new term 'plastisphere' describes the coexistence of microorganisms and plastic particles (Atlantik Technological University, 2022). Uncontrollable biofilms will be the result. It is possible that the plastic in people's bodies is already one of the reasons why many antibiotics are no longer effective (Zhu *et al.*, 2023). The dreaded septicaemia (blood poisoning) will once again lead to unnecessary deaths, as it did before the age of antibiotics. This issue is particularly feared by patients who need organ transplants, and by those who need long-term cannulas or other stomas such as chemo ports or anus praeter. We might need antiseptic essential oils to fix this increasing problem.

Preliminary findings also show that the tiny plastic particles are increasingly confusing our immune system. Our defences will therefore be gradually weakened, as researcher Berit Granum explained at the International

Conference on Microplastics, Nanoplastics & Human Health in Dublin in 2024 (Snapkow *et al.*, 2024). Cezmi Akdis from the University Zurich reported that the damage to the epithelial barriers caused by microplastics (they act like sandpaper on our outer skin and on the 'inner skins') leads to chronic inflammation (often unrecognized; Celebi Sozener *et al.*, 2022b; Celebi Sozener *et al.*, 2022a). Anti-inflammatory essential oils might contribute to a solution.

In Austria, it was shown that plastic particles are able to cross the blood–brain barrier and consequently also penetrate the brain, and that increased blood pressure (hypertension) may be even (partly) caused by plastic particles in the blood vessels (Geppner *et al.*, 2024). Antihypertensive essential oils might help affected patients.

A study conducted by Beijing Capital University investigated whether microplastics are present in the human heart and surrounding tissue. Tissue samples from various parts of the hearts of 15 patients were taken during open-heart surgery. Although microplastics were not consistently present in all tissue samples, nine types were found in five types of tissue, with the largest having a diameter of 469 µm (Yang *et al.*, 2023).

According to findings from a recent tissue study at the University of Vienna, microplastic particles can increase metastasizing cancer: they were transferred between cells during cell division. 0.25 µm plastic particles increase the tendency for cell migration and can have a positive effect on metastasis (Brynzak-Schreiber *et al.*, 2024). Essential oils which alleviate the harsh modern cancer treatments might be useful.

Many of these effects could be reduced through the targeted use of selected essential oils, as it can be assumed that neither the pollution nor the respective complaints can be remedied by conventional methods.

5.7   Conclusion

Although only a very small insight into clinical aromatherapy could be given here, these examples of rather serious health issues show that they could be improved or even resolved with the help of specific essential oils. They show that clinical aromatherapy can be an excellent tool, not only to improve many patients' health, but also to reduce the costs for many examinations, surgery and prolonged hospital stays. In order to remain transparent, it must be pointed out that many efficacy studies on EOs and their bioactive constituents have so far been based on *in vitro* tests. Unfortunately, complex interactions in living organisms have not yet been sufficiently investigated. This is why we often must rely on reports of experience, which are particularly valuable. However, this means that to remain viable for the future, we need to increase *in vivo* studies and clinical research and continue reports on case studies as valuable pieces of evidence. This could bring integrative therapeutic approaches and the use of authentic, genuine EOs a long way forward for patient's needs.

6

Use of Essential Oils in the Cosmetics Industry

Constanze Stiefel*

6.1 Demand of Essentials Oils for Cosmetic Purposes

The global market value of essential oils (EOs) was estimated to be nearly US$21 billion in 2022, with a market demand of approximately 300 kt of EOs (Fig. 6.1) and an expectation of further growth. According to the latest market data, the highest global market share is due to the application of EOs in food and beverages (about 38%), followed by fragrances, aromatherapy and cosmetics (about 29%), and household (16%). Other important applications include pharmaceuticals and animal feed (Sharmeen *et al.*, 2021; Frix, 2023).

For thousands of years, plant extracts, fatty oils and waxes, as well as EOs, have been used to enhance external beauty, protect and embellish the skin and improve overall health and physical appearance (McMullen and Dell'Acqua, 2023). Perfumes, in particular, played a vital role in most ancient civilizations (Stefania *et al.*, 2017). However, after the widespread use of synthetic and chemically modified cosmetic ingredients in the 20th century, it was not until the 1990s and early 2000s that there was a resurgence of interest in natural ingredients, driven in part by society's growing awareness of sustainable practices, individual well-being and health (Mahesh *et al.*, 2019). Many consumers perceive natural products as being healthier and having a lower environmental impact than synthetic products (NATRUE, 2021; Suphasomboon and Vassanadumrongdee, 2022). Accordingly, the cosmetic industry expects significant gains in terms of demand for natural ingredients, including EOs (Barbieri and Borsotto, 2018).

An estimated 3000 plants are known to produce EOs, but only 150–300 of these EOs are of commercial importance (Carvalho *et al.*, 2016; CBI, 2024). Among these, floral EOs from rose, gardenia, jasmine, lavender and ylang-ylang are still very popular ingredients for the cosmetic industry (Aburjai and Natsheh, 2003), while other EOs such as rosemary, tea tree, lemon,

*Corresponding author: constanze.stiefel@hs-esslingen.de

93

Fig. 6.1. Essential oils: Main segments of origin (Metric tons MT). (According to Frix, 2023; Author's own image using data from Frix and photos courtesy of WALA Heilmittel GmbH.)

orange, bergamot, ginger, chamomile, vetiver, peppermint, neroli, rosewood, sandalwood, cinnamon, thyme, clove, eucalyptus, patchouli, cedarwood, sage and vanilla are also commonly used (Manion and Widder, 2017; Sarkic and Stappen, 2018).

6.2   Definitions and Categorization of Scented Cosmetic Products

Cosmetic products in Europe must comply with the European Cosmetics Regulation (CPR) 1223/2009. The regulation defines a cosmetic product as 'any substance or mixture intended to be placed in contact with the external parts of the human body (epidermis, hair system, nails, lips and external genital organs) or with the teeth and the mucous membranes of the oral cavity with a view exclusively or mainly to cleaning them, perfuming them, changing their appearance, protecting them, keeping them in good condition or correcting body odours' (European Commission, 2009). In addition, although not clearly defined, terms such as 'cosmeceuticals' (a combination of cosmetics and pharmaceuticals) or 'medical skincare products' are also commonly used, referring to products containing special ingredients designed to address specific skin conditions such as very dry and irritated skin. It is important to note that while these products offer targeted skincare benefits, they are not classified as drugs and should not be considered a substitute for medical treatment of serious skin conditions (Kerscher and Buntrock, 2011). In addition, there is sometimes a fine line between aromatherapy and cosmetics. In particular, scented shower and bath products and massage oils tend to claim to have a distinct effect on the psyche and mood, like being revitalizing, activating or calming.

Cosmetics Europe, the European trade association for the cosmetics and personal care industry, divides cosmetic products into seven main categories – skincare, body care, hair care, sun care, oral care, decorative cosmetics, and perfumes – with further subdivisions (Fig. 6.2). Almost all of the cosmetic products mentioned are perfumed and thus contribute to the total daily exposure of the consumers.

Essential oils vary greatly in their price, depending on the amount of the EO in the plant and the effort required to extract it. For example, lavender oil is a relatively inexpensive oil with an average price of about 100 €/kg, while rose oil can cost over 10,000 €/kg (Fortineau, 2004). For cost reasons, the perfume industry also uses combinations of synthetic fragrances (fragrance or perfume oils) instead of EO blends, the former being produced by chemical synthesis, such as vanillin produced from lignin. In addition, 'semi-synthetic' fragrances are used, which are obtained by isolation (extraction and distillation) from a natural extract or EO, which may be further chemically modified (derivative). An example is eugenol obtained from clove oil and its derivative isoeugenol. However, synthetic fragrances that mimic an EO, consisting of several hundred individual components, by combining a few selected fragrance compounds may not contain the beneficial aspects of natural EOs and may never achieve the same olfactory complexity and depth.

6.3   Regulatory Framework

Before a cosmetic product can be placed on the European market, the manufacturer must ensure compliance with the European Cosmetics Regulation. As a result, the manufacturer or the responsible person needs to conduct a comprehensive safety assessment of the cosmetic product to ensure that it meets the safety requirements outlined in the regulation and prepare a Cosmetic Product Safety Report (CPSR). This involves assessing the safety

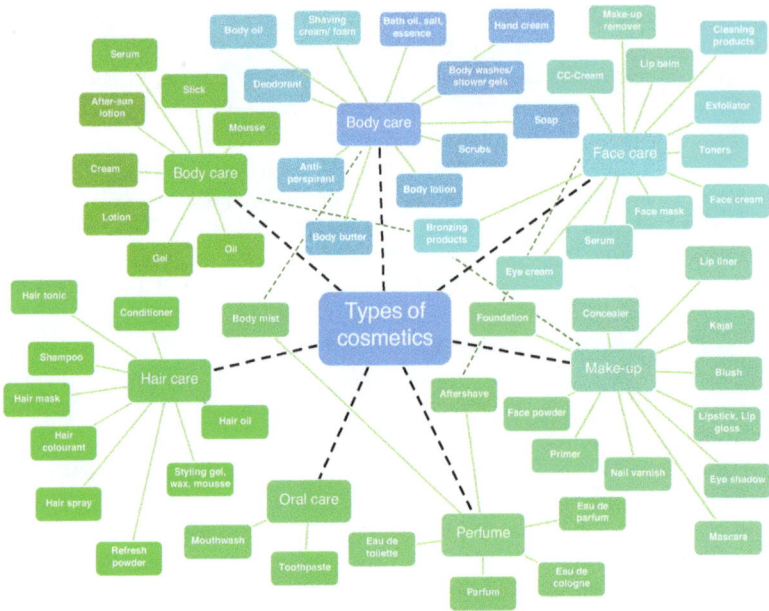

Fig. 6.2. Overview of different cosmetic product categories that can contain scented products. (According to Cosmetics Europe, 2017. Image adapted.)

of all ingredients used in the product formulation, including fragrances and EOs, under foreseeable conditions of use. It considers their concentrations in the product, the type of cosmetic product, the application site and the amount of product applied, the target population and the quality of the raw materials. Prohibited substances, listed in Annex II of the regulation (e.g. the EO of *Juniperus sabina*), must be avoided, and the requirements for restricted substances listed in Annex III (e.g. a maximum peroxide value of <10 mmol/L for *Pinus sylvestris* oil) must be met.

Furthermore, cosmetic manufacturers must ensure that the cosmetic ingredients they use comply with all relevant requirements of the European Chemicals Regulation REACH (1907/2006) and the CLP Regulation (1272/2008). According to Article 15 of the CPR, which refers to the CLP regulation, the use of CMR (carcinogenic, mutagenic and reprotoxic) substances is generally prohibited, except in certain cases where a substance classified as CMR 2 (suspected CMR) has been evaluated as safe by the Scientific Committee on Consumer Safety (SCCS) under specific conditions of use. Hence, a new classification for a single fragrance component can also affect the use of a natural EO in which it is present. A recent example is methyl salicylate, a major component of wintergreen and sweet birch EOs. Its CLP classification as a CMR substance of category 2 (toxic for reproduction) and the assessment of the SCCS led to its inclusion in Annex III of the Cosmetics Regulation, associated with strict maximum use levels (Scientific Committee on Consumer Safety, 2021; European Commission, 2022).

Additionally, specific national recommendations should be considered. For example, the German Federal Institute for Risk Assessment recommends maximum use concentrations of 1% for tea tree oil, camphor, eucalyptus oil and menthol in cosmetic leave-on products (Bundesinstitut für Risikobewertung, BfR, 2003, 2008).

For scented cosmetics, the International Fragrance Association (IFRA) plays a key role in consumer safety as the global representative body for the fragrance industry. Its primary mission is to promote the safe use of fragrance materials in consumer products, including cosmetics, personal care products and household goods. To accomplish this mission, IFRA establishes standards and guidelines that restrict or prohibit the use of specific fragrance materials, based on research conducted by the independent scientific Research Institute of Fragrance Materials (RIFM). Compliance with IFRA standards is mandatory for all IFRA members, which account for approximately 80% of the global production volume of fragrances. Due to their close collaboration with regulatory authorities, the IFRA standards are widely recognized as essential benchmarks to produce perfumed cosmetics.

In addition to product information requirements, it is important to ensure that product labelling complies with the requirements of the Cosmetics Regulation. This includes providing information such as the product name, list of ingredients (INCI; International Nomenclature of Cosmetic Ingredients), batch number, manufacturer's contact information and any necessary warnings or precautions.

6.4   Functions of Essential Oils in Cosmetics

Essential oils are widely used as ingredients in modern cosmetic products. One of the main reasons for their use is their pleasant fragrance:

- to mask unpleasant odours of other product ingredients (Sarkic and Stappen, 2018);
- to promote the purchase decision and product experience (Cuesta *et al.*, 2020);
- to create a signature fragrance impression of a brand (Kontaris *et al.*, 2020; Parente and Ares, 2021); and
- to serve as marketing advantage in sales and promotion (Salsabila, 2023).

The concentration of EOs in cosmetic products can vary widely, depending on the product type, intended use and desired fragrance strength. High-quality perfumes usually contain up to 20–30% EOs, while products like body lotions and face creams usually contain far less than 1% fragrance components. In particular, products for sensitive skin or medical skincare products are often completely unscented (Cadby *et al.*, 2002).

Besides their sensory qualities, EOs are known for their stabilizing and preserving properties and can enhance the stability of cosmetic products against bacteria and fungi (Herman *et al.*, 2013; Halla *et al.*, 2018). Although the use of chemical preservatives such as parabens is an effective way to protect cosmetic products from spoilage, their safe use is repeatedly questioned by consumers (Hansen *et al.*, 2012). Hence, there is keen interest in producing 'preservative-free' or 'self-preserving' cosmetics, using EOs and other plant extracts instead because of their pronounced antimicrobial potential (Bello *et al.*, 2022). Different EOs (e.g. lavender, tea tree, cinnamon, thyme, oregano, sage, lesser calamint, honeysuckle, rosemary) have been successfully evaluated for their antimicrobial and antifungal activities (Muyima *et al.*, 2002; Varvaresou *et al.*, 2009; Herman, 2014). However, the efficacy of EOs as accompanying preservatives can vary depending on the type and concentration of the oils used, the way they are incorporated into the formulation and the overall composition of the cosmetic product (Maccioni *et al.*, 2002; Cimino *et al.*, 2021).

Furthermore, EOs can offer a wide range of applications in cosmetics, see Table 6.1.

6.5   Potential Safety Issues of EOs Used in Cosmetics

As cosmetic products are intended to come into direct contact with the skin, the toxicological and dermatological safety of fragrances, as of all other ingredients, is of particular importance. Besides their positive effects and their centuries of use, EOs and individual fragrance components are also associated with negative aspects such as skin sensitizing and allergenic properties, phototoxicity or systemic effects, mostly related to individual components of the EOs.

6.5.1   Absorption and metabolism of fragrances in the skin

Systemic effects, in particular, depend crucially on the absorption of a substance into the body. The extent to which fragrance components are absorbed by the skin depends on a number of factors. First of all, absorption depends primarily on the lipophilicity of the fragrance but also on other properties such as its octanol-water partition coefficient, its molecular weight and its volatility, the dose and concentration applied, the surface area and region of application and the skin condition (Hotchkiss, 1998). Occlusive application conditions and the presence of other absorption-enhancing ingredients ('vehicles' such as ethanol) can increase the absorption rate (Brain *et al.*, 2022). On the other hand, the amount that can be absorbed is reduced by evaporation, perspiration,

Table 6.1. Potential proposed applications of some essential oils in different cosmetic categories. (Author's own table.)

Type of cosmetic product	Essential oils	Plant	Reference
Anti-ageing products	Patchouli Myrrh Frankincense Citronella Chamomile Rose Neroli and bitter orange peel Sandalwood Yarrow Geranium Lavender	*Pogostemon cablin* *Commiphora myrrha* *Boswellia serrata* *Cymbopogon nardus* *Matricaria chamomilla* *Rosa × damascena* *Citrus aurantium amara* *Santalum album* *Achillea millefolium* *Pelargonium graveolens* *Lavandula augustifolia*	(Charles Dorni *et al.*, 2017; Nayebi *et al.*, 2017; Francois-Newton *et al.*, 2021; Guzmán and Lucia, 2021; Rahmi *et al.*, 2021; Alraddadi and Shin, 2022; Ande and Bakal, 2022)
Anti-acne products	Tea tree Citronella Palmarosa Eucalyptus Myrtle Lavender Geranium Myrrh Clary sage Sandalwood Thyme Ylang-ylang	*Melaleuca alternifolia* *Cymbopogon nardus* *Cymbopogan martini* *Eucalyptus globulus* *Myrtus communis* *Lavandula angustifolia* *Pelargonium graveolens* *Commiphora myrrha* *Salvia sclarea* *Santalum album* *Thymus vulgaris* *Cananga odorata*	(Aburjai and Natsheh, 2003; Lertsatitthanakorn *et al.*, 2006; Nurzyńska-Wierdak *et al.*, 2022)
Toothpastes and mouthwashes	Sage Peppermint Corn mint Eucalyptus Lemon Myrrh Clove Lemon balm	*Salvia officinalis* *Mentha × piperita* *Mentha arvensis* *Eucalyptus globulus* *Citrus limon* *Commiphora myrrha* *Eugenia caryophyllus* *Melissa officinalis*	(Smolarek *et al.*, 2015; Mazur *et al.*, 2022)
Deodorants	Sage Rosemary Thyme Lavender Peppermint Lemon Rose	*Salvia officinalis* *Salvia rosmarinus* *Thymus vulgaris* *Lavendula officinalis* *Mentha × piperita* *Citrus limon* *Rosa × damascena*	(Oliveira *et al.*, 2021)

Continued

Table 6.1. Continued

Type of cosmetic product	Essential oils	Plant	Reference
After-sun products	Lavender	*Lavandula augustifolia*	(Carvalho *et al.*, 2016; Arraiza, 2017)
	Chamomile	*Matricaria chamomilla*	
	Peppermint	*Mentha × piperita*	
Anti-dandruff shampoos	Tea tree	*Melaleuca alternifolia*	(Abelan *et al.*, 2022; Jain *et al.*, 2022)
	Rosemary	*Salvia rosmarinus*	
	Lemon grass	*Cymbopogon citratus*	
	Bergamot	*Citrus bergamia*	
	Cinnamon	*Cinnamomum verum*	
	Thyme	*Thymus vulgaris*	
Hair growth stimulant	Bergamot	*Citrus bergamia*	(Abelan *et al.*, 2022)
	Rosemary	*Salvia rosmarinus*	
	Chamomile	*Matricaria chamomilla*	
	Peppermint	*Mentha × piperita*	

washing, abrasion and bacterial degradation. In addition, the skin can act as a reservoir for many fragrances, from which they are slowly released into the systemic circulation (Hotchkiss, 1998).

Furthermore, the skin is a highly metabolically active tissue that contains several enzymes. These enzymes catalyse reactions such as oxidation, reduction or hydrolysis, as well as conjugation reactions that allow the substances to be transported away in the blood. For example, in humans, benzyl acetate is almost completely (99%) hydrolysed to benzyl alcohol in the skin (Hotchkiss, 1998). In contrast, other substances (e.g. coumarin) are not degraded in the skin upon absorption (Yourick and Bronaugh, 1997). Conversion and degradation reactions may also be of toxicological significance as they alter the irritating or sensitizing effects of, for example, reactive aldehydes such as cinnamaldehyde and hydroxycitronellal. If a fragrance itself is non- or low-sensitizing, but can be converted into a hapten in the skin (bioactivation), it is called a prohapten (Karlberg *et al.*, 2013).

6.5.2 Alkylbenzenes safrole, methyleugenol, isoeugenol and estragole

Due to studies in rodents indicating potential carcinogenic effects in the liver following oral administration of high doses of the pure alkenylbenzenes estragole, methyleugenol, isoeugenol and safrole, these components have been restricted for cosmetic use as a precautionary measure, as some absorption through the skin cannot be excluded. Methyleugenol is typically found in authentic EOs of citronella, basil, bay laurel and tea tree; isoeugenol is a natural component of perilla, nutmeg, basil, cinnamon, clove, ylang-ylang and ginger. EOs of estragon, basil, pine, turpentine, fennel, rose and anise contain estragole in partly high concentrations, while safrole is a typical constituent of anise, camphor, nutmeg, cinnamon and black pepper. When such oils are used in cosmetic products, the Cosmetics Regulation stipulates that the maximum concentration of methyleugenol in the finished products must not exceed 0.01% in a fine fragrance, 0.004% in eau de toilette, 0.002% in a fragrance cream, 0.0002% in other leave-on products and in oral hygiene products as well as 0.001% in rinse-off products. Safrole must not exceed 0.01% in the finished product and 0.005% in products for dental and oral hygiene, while isoeugenol should not exceed 0.02%. For estragole, there is an IFRA standard that defines maximum concentrations between 0.00021 and 0.014% in the finished cosmetic product, depending on the product category (International Fragrance Association, 2023).

6.5.3   Phototoxicity due to furocoumarins

Furocoumarins are a well-known group of natural phototoxins, found primarily in citrus peel oils, but also in EOs of parsley leaf, angelica root, marigold, coriander, carrot and rue, with bergapten (5-methoxypsoralen, 5-MOP), bergamottin and psoralen being the most common (Cohen *et al.*, 2019). When furanocoumarins come into contact with the skin and are subsequently exposed to sunlight (mainly UVA radiation), burn-like symptoms may occur, including redness of the skin, swelling, blistering, lesions and photopigmentation. Since this reaction is concentration dependent (Kejlová *et al.*, 2010), IFRA concluded on a limit of 5 ppm in leave-on products for any combination of seven marker furocoumarins (bergapten, bergamottin, byakangelicol, epoxybergamottin, isopimpinellin, oxypeucedanin, xanthotoxin) and a maximum of 1 ppm for each component in 2008. For rinse-off products, a limit of 50 ppm was proposed (International Fragrance Association, 2008). However, the topic is still under discussion. For instance, in 2015 IFRA stipulated that the maximum acceptable concentration of 5-MOP in leave-on products must not exceed 0.0015% (15 ppm), while no restrictions for rinse-off products were included (International Fragrance Association, 2023). In addition, the Cosmetics Regulation requires that the furocoumarin content in sun-protection and bronzing products has to be below 1 ppm, and in Switzerland, a 1 ppm limit applies to all leave-on products that may be exposed to the sun. In contrast, a recent study demonstrated the absence of phototoxicity and a photoirritation potential of bergamottin (Cluzel *et al.*, 2022).

    To meet those limits, EOs containing furocoumarins are often processed in ways that reduce their furocoumarin content (Gionfriddo *et al.*, 2004; Valussi *et al.*, 2021). However, since every single fragrance contributes to the overall scent impression of an EO, such treatments inevitably affect the olfactory quality of the oil. Additionally, it has been shown that furocoumarins also have a UV-protective effect on further EO components such as limonene and γ-terpinene, preventing the formation of unwanted oxidation products and off-flavours (Bitterling *et al.*, 2022).

6.5.4 Oxidation of essential oils and fragrances

Although some EOs have been identified as contact allergens, sensitization occurs only infrequently (Sabroe *et al.*, 2016; Geier *et al.*, 2022). Also, some fragrance components such as limonene, linalool, geraniol, geranial, α-terpinene or linalyl acetate have a relatively low allergenic potential per se. However, their molecular structure, which contains oxidizable allylic positions, predisposes them to the formation of reactive oxidation products such as hydroperoxides and other secondary oxidation products such as epoxides or aldehydes upon exposure to air (Turek and Stintzing, 2013). These products are closely associated with an increased incidence of allergic reactions and are therefore referred to as prehaptens (Karlberg *et al.*, 2013). The same applies to some oxidized EOs, such as lavender or tea tree oil, which also contain some of the fragrances mentioned (Barbaud *et al.*, 2023). Heat, light exposure and catalytically active impurities such as metal ions can further enhance these oxidation processes (Bitterling *et al.*, 2020). However, the clinical relevance of these oxidation reactions is questionable since an increased formation of oxidation products in cosmetics themselves has not yet been confirmed. For example, Natsch *et al.* found no evidence for oxidation during the storage of several tested cosmetic products (Natsch *et al.*, 2019).

Different measures can be taken to stabilize terpenoids and other sensitive components in the EO itself or in the final cosmetic products. For instance, proper handling and storage of EOs (controlled temperature, inert gas atmosphere, firmly sealed, light protection) can minimize the risk of adverse skin reactions and preserve their efficacy (Turek and Stintzing, 2012). In addition, in a complex cosmetic formulation, other antioxidant ingredients may have a stabilizing effect on rather vulnerable ingredients. Furthermore, encapsulation techniques such as coacervation, spray-drying, extrusion, fluid bed coating, solvent evaporation, ionic gelation or liposomal encapsulation (see Chapter 3) can effectively prevent or minimize degradation processes (Carvalho *et al.*, 2016; Sousa *et al.*, 2022). Moreover, encapsulation can also improve the incorporation of EOs into cosmetic formulation and can modulate their release (Yammine *et al.*, 2024).

6.5.5 Contact allergy

Substances known to cause widespread allergic reactions are not used in cosmetics. And most EOs used in cosmetics are considered well-tolerable when used in high quality and in accordance with regulatory guidelines. However, as described above, certain EOs may be more likely to cause adverse reactions due to specific fragrance components or increased reactivity.

Several studies have shown that around 1–3% of the total European population may be affected by allergic contact dermatitis (ACD) caused by single fragrance components (Johansen, 2003; Diepgen *et al.*, 2016; Geier *et al.*, 2022). Typical symptoms of ACD include redness and swelling, inflammation and dryness of the skin and formation of itchy blisters. ACD is a type IV immunologic reaction involving the T lymphocytes of the immune system (Martin *et al.*, 2018b). Most skin reactions to cosmetics are thought to be irritant, while it is estimated that less than 10% of the observed reactions are contact allergies (De Groot *et al.*, 1988).

6.6 Diagnosis of Existing Sensitizations in Consumers

Already since the late 1970s, contact sensitization to fragrances has been screened by patch testing of a standard fragrance mixture (FM I) consisting of amyl cinnamal, cinnamal, cinnamyl alcohol, eugenol, geraniol, hydroxy-citronellal, isoeugenol and oakmoss absolute (*Evernia prunastri*) at relatively high concentrations of overall 8% and further a 5% of sorbitan sesquioleate as emulsifier, which has a certain inherent sensitization potential (Geier *et al.*, 2015a). In 2005, a second mixture (FM II) was established, containing citronellol, coumarin, farnesol, hexyl cinnamaldehyde and hydroxyisohexyl 3-cyclohexene carboxaldehyde (HICC, Lyral) at a total concentration of 14% in petrolatum. More recently, oxidized fragrance allergens from linalool and limonene have become available as clinical test substances and are frequently tested by default (Schnuch and Griem, 2018). Initial clinical data confirm the already mentioned higher prevalence of allergic reactions to the degraded, oxidized fragrances as opposed to the original, unaltered fragrance, confirming the importance of proper protection against oxidation (Ogueta *et al.*, 2022; Schubert *et al.*, 2023).

In highly sensitive patients with suspected ACD, typically tested with FM I and II, the prevalence of fragrance allergy has been estimated to be between 6 and 14% (Geier *et al.*, 2015b). Of course, the data obtained from dermatitis patients cannot simply be extrapolated to the general population, as this is a selected group of people with specific skin physiological conditions.

Dermatological data show that the prevalence of clear allergic reactions to EOs in the same patient group is significantly lower at 0.2–2.5%, while test concentrations of up to 10% were used, which are significantly higher than those commonly found in cosmetics (Geier *et al.*, 2022). Accordingly, a positive reaction to fragrance mix I or II does not automatically mean that a person must avoid cosmetic products containing EOs. This statement was confirmed by a study that quantitatively recorded the skin tolerance to EOs in patients with a documented contact allergy as determined by the standard patch test. The patients showed a significant difference between their tolerance to the fragrance mix and their tolerance to the products containing the natural EOs (Meyer, 2004).

6.7   Safety of Scented Cosmetic Products

Because the quality of EOs is critical for their safe use, cosmetic manufacturers adhere to strict quality parameters outlined in raw material specifications for every EO. Only when the specifications are met the raw material is released for the production process. To ensure the safety and consistent quality of every cosmetic product, a combination of measures is applied (Fig. 6.3).

In addition to all the precautionary measures, a comprehensive safety assessment for every cosmetic product and proper ingredient labelling can further ensure safe cosmetic usage (Bialas *et al.*, 2023). As part of the safety assessment and in accordance with the IFRA standards, the concept of quantitative risk assessment (QRA) is used. Within this concept, safe concentration thresholds (no expected sensitization induction level, NESIL) for fragrances and EOs are established, below which a fragrance ingredient is not expected to induce skin sensitization in the general population. These NESIL values can be derived from animal data (local lymph node assay, LLNA, and sensitization potency expressed as an EC3 value) or from alternative methods recognized by the Organisation for Economic Co-operation and Development (OECD). Using the NESIL in combination with different sensitization assessment factors, which consider potency, exposure levels and frequency of use, acceptable exposure levels (AEL) are derived (Api *et al.*, 2008; Api *et al.*, 2020). By adhering to these exposure levels, cosmetic manufacturers ensure safe use of their products.

6.7.1   Labelling requirements

Under the Cosmetics Regulation, ingredients are labelled according to the International Nomenclature of Cosmetic Ingredients (INCI). Since fragrances are complex mixtures of various individual substances, fragrances or flavours (e.g. in oral care products) are generally declared under the collective term 'perfume'. Due to their possible sensitizing potential, the Cosmetics Regulation additionally requires the declaration of certain EOs and fragrances in the list of ingredients, if they exceed a concentration of 0.01% in rinse-off cosmetics (e.g. soap, shower gel, shampoo) or 0.001% in leave-on products (e.g. cream, lotion, tonic), regardless of whether the fragrance components are used as isolated substances or as part of an EO. Since the 26th amendment of the EU Cosmetics Regulation in 2012, 26 fragrance substances (24 single components and 2 plant extracts) had to be declared, including the fragrances tested with FM I and II. Although it is widely recognized that contact allergies to some of the 26 fragrance ingredients that were originally required to be declared were highly prevalent (e.g. oak moss, lyral or isoeugenol), while

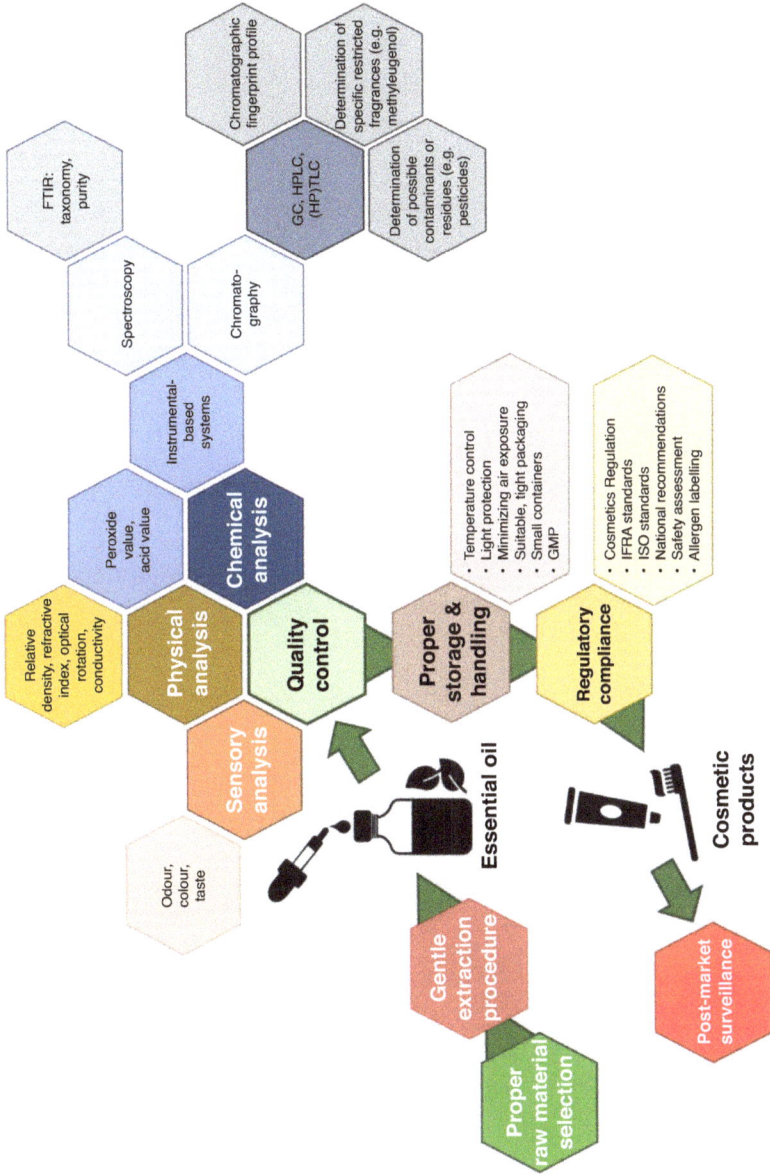

Fig. 6.3. Measures to ensure quality and safety of essential oils and the cosmetic products in which they are used. (By Constanze Stiefel.)

reactions to other ingredients (e.g. amyl cinnamal, coumarin or citronellol) were much rarer and of low clinical relevance (Geier *et al.*, 2015b; Schnuch and Griem, 2018), the European Commission decided to extend the labelling requirements by 56 fragrances and EOs in 2023, based on the assessment of the SCCS on fragrance allergens in 2012 (Scientific Committee on Consumer Safety, 2012). 'New allergens' include, for example, menthol, terpineol (alpha- and mixture of isomers), linalyl acetate, camphor and vanillin, as well as well-known EOs, such as cinnamon and lavender oil (European Commission, 2023). It is doubtful whether this expansion will be able to contribute to the stated goal of further consumer protection, as long as standard test materials for clinical patch testing are not available for a large proportion of the newly added substances.

6.7.2   Market surveillance

A proper documentation of all reported, plausible intolerance reactions to any cosmetic product is essential, and therefore functional market surveillance is a basic requirement for all cosmetic manufacturers in Europe. In this way, unexpected intolerances can be recognised at an early stage and, if possible, traced back to specific ingredients. In Germany, these data are collected and evaluated by the German Cosmetic, Toiletry, Perfumery and Detergent Association (IKW). In the years 2006–2021, only 31 medically confirmed (and a total of 60) intolerances were registered in over 28 billion cosmetic products sold. This confirms that the measures mentioned above are effective and that cosmetic products are generally well tolerated (IKW, 2023).

7 Use of Essential Oils and Plant Extracts in the Food Industry

Hartwig Schulz* and Michael Wink

7.1 History of the Most Important Aromatic and Spice Plants

Essential oils (EOs) have been obtained from plant material for thousands of years. Archaeological findings show that the technique of distillation was already used to produce fragrances in Persia in the 3rd millennium BC (Day, 2013). One of the oldest medical textbooks from 2000 BC, *The Yellow Emperor's Book of Internal Medicine*, comes from China and was written by the 'Yellow Emperor' Huang Ti (Ni, 1995). At that time, the Egyptians also knew about the disinfecting effect of EOs and used them mainly for cosmetics and mummification. Around 1000 AD, the distillation technique was rediscovered by the Arabs and in the 12th century this knowledge reached Europe when the Arabs founded the first universities in Spain and southern France. Even in the Roman Empire, spices from exotic countries were used to flavour food, both to emphasize its typical taste and to mask the bad taste of meat due to the onset of spoilage. In this context, Pliny the Elder wrote a treatise on 'Natural History' at this time, which describes various observations on aromatic plants and their individual uses (Pybus, 2006).

At the end of the 12th century, there existed already the first perfumers in Europe working with EOs. Later, the distillation book of the Strasbourg physician Brunschwig described the extraction of EOs (Eamon, 2000). In 1598, the Nuremberg and Augsburg pharmacopoeias documented the use of 108 distilled EOs in trade and commerce (Taape, 2014), which were mainly used in medicine and perfumery. Although steam distillation was already successfully used in ancient times to extract various EOs from different parts of plants (flowers, leaves, fruits, bark), it was not until the end of the 19th century that

*Corresponding author: hs.consulting.map@t-online.de

it was recognized that the products obtained in this way could also be used to flavour various foods (Rothe, 1988).

In the second half of the 19th century, the first flavour and fragrance companies emerged, particularly in Great Britain and Germany. In line with the rapid development of organic chemistry, the focus was primarily on identification, large-scale synthesis and commercialization, thus reducing the sometimes high costs of imported plant-based raw materials (Rowe, 2006).

Depending on the individual genotype (variety), geographical origin, cultivation conditions, plant parts used and the extraction/distillation process applied, the various industrially processed plant species often yield very different ingredient profiles with different organoleptic and efficacy profiles. Some selected plant species and their characteristic compounds are listed below in Tables 7.1 and 7.2, as well as in Fig. 7.1. Interestingly, the characteristic compounds are not single ones, but either volatiles or both volatiles and non-volatiles, making them a natural complex substance (NCS).

7.2 Production of Various Food Flavourings

7.2.1 Biotechnology

The history of biotechnology dates back to 6000 BC, when the Sumerians and Babylonians had already developed the first microbial processes for the production of beer, wine and bread. In the past 150 years, however, research and development in this context has focused primarily on the production of pharmaceuticals, chemical raw materials and flavourings, which can then be declared as natural. Intact microbial cells, crude enzyme mixtures or isolated enzymes are mainly used for the biotechnological production of natural flavours or to improve the individual flavour impression. The enzymes/enzyme classes employed to bioconvert certain raw materials into specific flavour substances are listed in Table 7.3 (Janssens et al., 1992; Rowe, 2006).

7.2.2 Reaction flavourings

When food is processed, especially when amino acids/proteins and sugars are heated, the so-called Maillard reaction significantly changes the taste and colour of individual foods. Today, this natural process is used on a large scale to produce various reaction flavours (especially meat flavours). Since Maillard reactions are influenced by various factors such as the protein source, hydrolysis conditions, molecular weight of the polypeptide, temperature and pH, attempts have been made in recent years to investigate the parameters relevant to the production of flavour-enhancing peptides in more detail in order to establish a better correlation with individual consumer preferences and acceptance. In addition, the existing negative effects of Maillard reaction on the biological properties of proteins have also been researched (Liu et al., 2022).

Table 7.1. Aromatic and aromatic-bitter tasting herbs and spices; EO = essential oil. From van Wyk and Wink, 2015, 2017. (Author's own table.)

Scientific name	Trivial name	Characteristic compounds	Sensory quality	Properties
Abies alba and related species	silver fir	EO: pinene, bornyl-acetate, phellandrene, camphene	aromatic	antiseptic, expectorant, counter-irritant
Aloysia citridora	vervain	EO: citral, limonene; flavonoids	aromatic	probably, sedative and anxiolytic properties
Alpinia officinarum	galangal, Siamese ginger	EO: non-volatile galangols and gingerols	aromatic, pungent	antispasmodic, anti-inflammatory, and antimicrobial
Anethum graveolens	dill	EO: *R*-(−)-carvone, limonene, dill ether	aromatic spice	antimicrobial and antispasmodic
Angelica archangelica	archangel, garden angelica	EO: furanocoumarins, coumarins	aromatic	digestive, antispasmodic, cholagogue; phototoxic and mutagenic
Apium graveolens	celery	EO: limonene, apiole, butylphthalides, methylphthalides	aromatic spice	sedative, spasmolytic
Artemisia dracunculus	tarragon	EO: estragol, phellandrene, ocimene	aromatic spice	appetite stimulant, mild sedative
Boswellia sacra and related species	frankincense	EO: α-pinene, phellandrene; triterpenoids (boswellic acid)	aromatic	antimicrobial, anti-inflammatory
Calendula officinalis	marigold	EO: α-cadinol, ionone; flavonoids, triterpenes	aromatic	anti-inflammatory, antispasmodic
Carum carvi	caraway	EO: *D*-(−)-carvone, limonene, carveol; phenylpropanoids	aromatic	antimicrobial, spasmolytic, carminative

Continued

Table 7.1. Continued

Scientific name	Trivial name	Characteristic compounds	Sensory quality	Properties
Cinnamomum camphora	camphor tree	EO: camphor, 1,8-cineole	aromatic	antiseptic, carminative, spasmolytic, counter-irritant
Cinnamomum verum, C. aromaticum	cinnamon bark tree	EO: cinnamaldehyde, eugenol	aromatic	antimicrobial, antispasmodic, choleretic
Citrus aurantium, Citrus paradisi and related species	bitter orange	EO: limonene, linalool, terpineol; bitter naringenin, neohesperidin	aromatic, bitter	stomachic, appetite stimulant
Commiphora myrrha	myrrh tree	EO: furanoeudesma-1,3-diene	aromatic	antimicrobial, anti-inflammatory, antipyretic
Coriandrum sativum	coriander	EO: linalool	aromatic	antimicrobial, spasmolytic, carminative
Crocus sativus	saffron	EO: safranal; crocetin, picrocrocin	aromatic	antispasmodic, sedative
Cymbopogon citratus	lemongrass	EO: geranial, neral	aromatic	antimicrobial, spasmolytic, carminative
Elettaria cardamomum	cardamom	EO: 1,8-cineole	aromatic	antimicrobial, spasmolytic
Eucalyptus globulus and related species	blue gum, eucalyptus	EO: 1,8,-cineole; sesquiterpenes, flavonoids	aromatic	antimicrobial, spasmolytic, expectorant
Filipendula ulmaria	meadowsweet	EO: methyl salicylate, salicylaldehyde; glycosides, tannins	aromatic	anti-inflammatory, analgesic, antirheumatic
Foeniculum vulgare	fennel	EO: trans-anethole, fenchone,	aromatic, bitter (fenchone)	carminative, spasmolytic, expectorant
Gaultheria procumbens	wintergreen	methyl salicylate; arbutin, tannins	aromatic	anti-inflammatory, analgesic, antirheumatic

Hyssopus officinalis	hyssop	EO: pinocamphone; marrubiin, rosmarinic acid	aromatic	anti-inflammatory, antispasmodic, antiseptic
Illicium verum	star anise	EO: *trans*-anethole	aromatic spice	carminative, antispasmodic, antiseptic, diuretic
Juniperus communis	juniper	EO: pinene, terpinen-4-ol	aromatic	diuretic, spasmolytic, antimicrobial
Laurus nobilis	bay	EO: 1,8-cineole, eugenol, linalool, alkaloids	aromatic	sedative, spasmolytic
Lavandula angustifolia	lavender	EO: linalyl acetate, linalool	aromatic	spasmolytic, carminative, sedative
Levisticum officinale	lovage	EO: alkylphthalides	aromatic	
Matricaria chamomilla	chamomile	EO: sesquiterpenes (α-bisabolol, chamazulene)	aromatic	anti-inflammatory, antispasmodic, carminative, antiseptic
Melaleuca alternifolia and related species	tea tree	EO: terpinen-4-ol, terpinene, 1,8-cineole	aromatic	antibacterial, antifungal
Melissa officinalis	lemon balm	EO: citronellal, citral A and B; rosmarinic acid, flavonoids	aromatic	spasmolytic, sedative, carminative, antiviral (rosmarinic acid)
Mentha × piperita and related species	peppermint	EO: menthol	aromatic	antimicrobial, analgesic, spasmolytic, carminative, choleretic
Myroxylon balsamum	tolu balsam tree	Oleoresin with benzoic and cinnamic acid and their esters	aromatic, fragrant	spasmolytic, vasodilating, antibacterial
Myrtus communis	myrtle	EO: 1,8-cineole, pinene, myrtenol, limonene	aromatic	antimicrobial, spasmolytic, expectorant

Continued

Table 7.1. Continued

Scientific name	Trivial name	Characteristic compounds	Sensory quality	Properties
Nigella sativa	black seed	EO: thymoquinone	aromatic, spicy	antispasmodic, diuretic
Ocimum basilicum	sweet basil	EO: methylchavicol, linalool, eugenol	aromatic	anti-inflammatory, antimicrobial
Ocimum tenuiflorum	holy basil	EO: eugenol, methyleugenol, caryophyllene	aromatic	anti-inflammatory, antimicrobial
Origanum majorana	marjoram	EO: sabinene, terpinene, terpinen-4-ol, 1,8-cineole	aromatic	anti-inflammatory, carminative, antiseptic
Origanum vulgare and related species	oregano	EO: carvacrol, *p*-cymene, terpinene	aromatic	anti-inflammatory, carminative, antiseptic
Orthosiphon aristatus	orthosiphon	EO: borneol, limonene, thymol; diterpenes, flavonoids	aromatic	antimicrobial, diuretic, anti-inflammatory, antioxidant
Petroselinum crispum	parsley	EO: phenylpropanoids such as apiol, myristicin	aromatic spice	diuretic, antipruritic
Picea abies	spruce	EO: pinene, bornyl acetate, phellandrene, camphene	aromatic	antiseptic, expectorant, counter-irritant
Pimpinella anisum	anise	EO: *trans*-anethole	aromatic	carminative, spasmolytic, antimicrobial, expectorant
Pimpinella major	greater burnet saxifrage	EO: epoxypseudoeugenol and its esters	aromatic	secretolytic, secretomotoric
Pinus sylvestris and related species	Scots pine	EO: pinene, bornyl acetate	aromatic	antiseptic, secretolytic, expectorant, counter-irritant
Pogostemon cablin	patchouli	EO: sesquiterpenes, patchoulil	aromatic	antimicrobial, skin protectant

Populus tremuloides	American aspen	benzoyl esters of salicin, salicin, salicortin	aromatic	the phenolics are converted to salicylic acid in the body and have potent anti-inflammatory analgesic properties
Salix alba and related species	white willow	salicin, salicortin	aromatic/bitter	the phenolics are converted to salicylic acid in the body and have potent anti-inflammatory analgesic properties
Salvia rosmarinus (formerly *Rosmarinus officinalis*)	rosemary	EO: 1,8-cineole, pinene, camphor; rosmarinic acid, carnosic acid	aromatic	antimicrobial, spasmolytic, choleretic, antioxidant
Salvia officinalis and related species	sage	EO: thujone, carnosic acid, rosmarinic acid	bitter aromatic	antiseptic, carminative, antispasmodic activity; thujone is a neurotoxin
Santalum album	sandalwood	EO: santalol	herbal spicy	antibacterial, spasmolytic
Satureja montana	savory	EO: carvacrol, p-cymol, terpinen-4-ol, thymol	aromatic spice	antimicrobial, diuretic, antispasmodic, secretomotoric

Continued

Table 7.1. Continued

Scientific name	Trivial name	Characteristic compounds	Sensory quality	Properties
Syzygium aromaticum	clove tree	EO: eugenol, caryophyllene	aromatic	local anesthetic, antiseptic, carminative
Tagetes minuta and related species	Mexican marigold	EO: tagetone, dihydrotagetone, ocimene,	aromatic	antiseptic, carminative, antispasmodic, anti-inflammatory
Thuja occidentalis	white cedar	EO: thujone	aromatic	diuretic, decongestant; thujone is a neurotoxin
Thymus vulgaris and related species	thyme	EO: thymol, carvacrol	aromatic	antiseptic, antispasmodic, expectorant
Vanilla planifolia	vanilla	vanillin (formed from the glycoside vanilloside)	aromatic	antioxidant, anti-inflammatory

Table 7.2. Herbs and spices with bitter and pungent taste; EO = essential oil. (Author's own table.)

Scientific name	Trivial name	Characteristic compounds	Sensory quality	Properties
Acorus calamus	calamus	EO: acorenone, camphene, *p*-cymene, linalool; β-asarone	spicy aromatic,	spasmolytic, CNS sedative; mutagenic (asarone)
Allium cepa, A. sativum	onion, garlic	sulfur-containing phytochemicals; alliin is converted to pungent allicin	pungent	antimicrobial, antiviral, lipid-lowering, hypotensive
Aloe vera, A. ferox	aloe	bitter anthraquinones, polysaccharides	bitter	strong laxative activity
Andrographis paniculata	king of bitters	bitter diterpene lactones (andrographolides)	bitter	antimicrobial, anti-inflammatory, respiratory tract infections
Armoracia rusticana	horseradish	Glucosinolates, which contain the pungent-tasting allyl isothiocyanates	pungent spice	antimicrobial, spasmolytic, hyperaemic
Artemisia absinthium	wormwood	EO: thujone, chrysanthenyl acetate; bitter sesquiterpene lactones	aromatic and bitter tonic	antimicrobial; psychoactive
Brassica nigra	black mustard	Glucosinolates, which contain the pungent-tasting allyl isothiocyanates	pungent spice	antimicrobial, spasmolytic, counter-irritants
Canella winterana	winter bark	EO: pinene, eugenol; bioreactive sesquiterpenes (dialdehydes): muzigadial	pungent (peppery taste)	antimicrobial, antifeedant
Capsicum frutescens, C. anuum	chili pepper	capsaicin and other capsinoids	pungent	topical analgesic, counter-irritant, carminative

Continued

Table 7.2. Continued

Scientific name	Trivial name	Characteristic compounds	Sensory quality	Properties
Cinchona pubescens	red cinchona	quinine, quinidine (quinoline alkaloids)	bitter	antimalarial; quinidine inhibits sodium channels
Curcuma longa	turmeric	EO: bisabolene; curcumin and other curcuminoids	bitter, aromatic	antimicrobial, anti-inflammatory, antioxidative
Ferula assa-foetida	asafoetida	disulfides, polysulfanes, sesquiterpenes	bitter, pungent	carminative, antispasmodic, expectorant
Gentiana lutea	yellow gentian	bitter secoiridoids, gentiopicroside, amarogentin	bitter	digestive, antimicrobial
Glycyrrhiza glabra, *G. uralensis*	liquorice	flavonoids (liquiritin); triterpene saponins (glycyrrhizic acid)	sweet	expectorant, secretolytic, anti-inflammatory
Humulus lupulus	hop	2-methyl-3-buten-2-ol	bitter	sedative
Marrubium vulgare	white horehound	EO: bitter diterpene lactones (marrubiin)	bitter	expectorant, choleretic
Myristica fragrans	nutmeg tree: fruit/fruit coverings (mace)	EO: sabinene, pinene, myristicin, elemicin, eugenol	aromatic, pungent	antimicrobial, anti-inflammatory; myristicin is a psychostimulant
Pimenta dioica	allspice	EO: eugenol, 1,8-cineole, caryophyllene	spicy, clove-like	antioxidant
Piper nigrum, *Piper cubeba,* *Piper longum* and related species	pepper	EO: piperine and related alkaloids, sabinene, pinene, phellandrene, linalool	aromatic, pungent	antimicrobial, cholagogue, digestive, stimulant
Schinus molle	pink pepper	EO: pinene, terpineol, phellandrene	aromatic, pungent	antimicrobial, anti-inflammatory, diuretic

Continued

Table 7.2. Continued

Scientific name	Trivial name	Characteristic compounds	Sensory quality	Properties
Sinapis alba	white mustard	Glucosinolates, which contain the pungent-tasting allyl isothiocyanates	pungent	antimicrobial, spasmolytic, counter-irritant
Stevia rebaudiana	stevia	diterpenoid glycosides: stevioside	sweet	sugar substitute
Trigonella foenum-graecum	fenugreek	steroid saponins, alkaloids	bitter	hypoglycaemic, cholesterol-lowering
Tropaeolum majus	nasturtium	glucosinolates; benzyl isothiocyanate	pungent	antimicrobial, spasmolytic, hyperaemic
Verbena officinalis	vervain	iridoid glycosides: verbenalin	bitter	diuretic, expectorant, anti-inflammatory
Warburgia salutaris	pepperbark tree	drimane sesquiterpenoids (warburganal, polygodial)	pungent	antibacterial, anti-inflammatory
Zingiber officinale	ginger	EO: camphene, phellandrene; sesquiterpenes (zingiberene), gingerol	pungent	antibacterial, anti-inflammatory, carminative, hypoglycaemic

7.2.3 Preservation of freshness and extending shelf life

The aromatic compounds in EOs can mask or reduce the development of off-flavours and odours associated with food spoilage, thus preserving the sensory quality of the food (for example various citrus EOs add a fresh and vibrant flavour while masking any undesirable tastes).

Essential oils can be effective in preventing the growth of bacteria, moulds and yeasts, which are common culprits in food spoilage, especially in baked goods and fruits. Some EOs, like citronella and lemongrass, possess antifungal properties that can help extend the shelf life of certain food products. They are used as natural alternatives to various synthetic preservatives and are incorporated into food formulations to replace or reduce the reliance on chemical preservatives. This aspect is particularly relevant in the context of consumer preferences for natural and 'clean-label' products. Nevertheless, it is important to note that while EOs offer natural preservation benefits, their effectiveness can vary dramatically

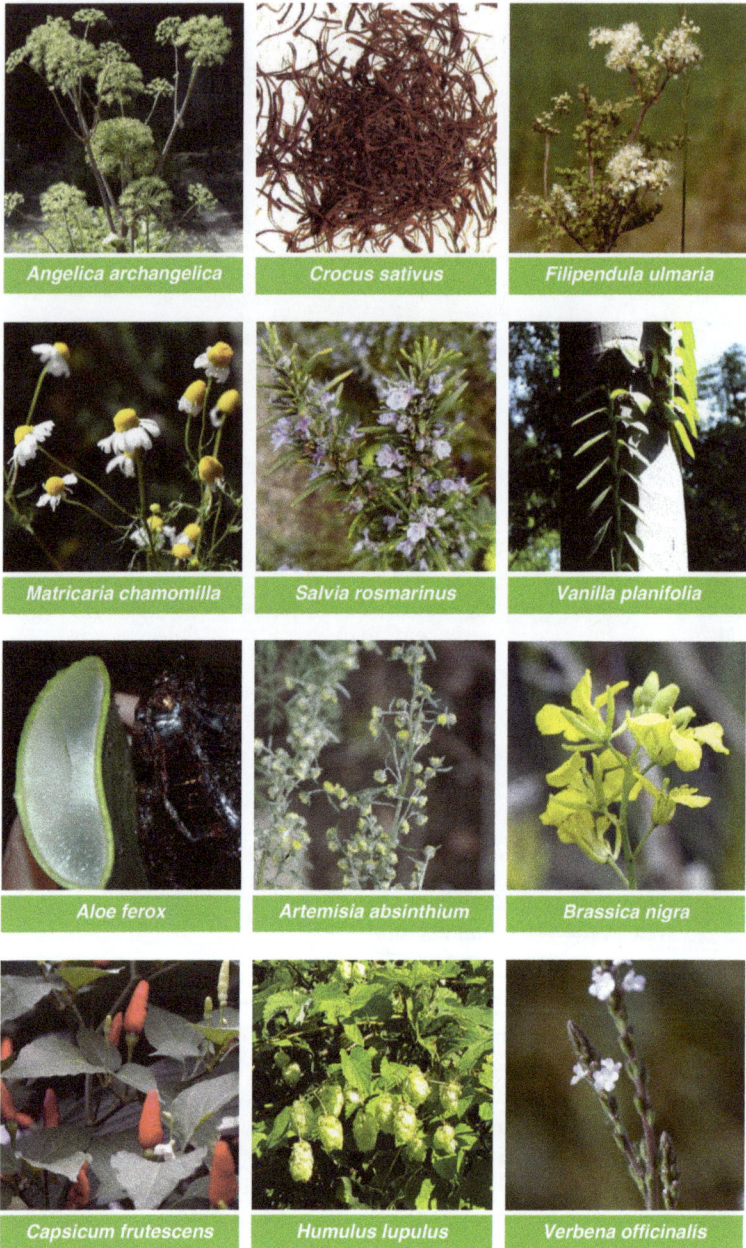

Fig. 7.1. Overview of some selected plant species with aromatic and aromatic-bitter taste (Table 7.1), as well as with bitter and pungent taste (Table 7.2). (Photos: Michael Wink.).

Table 7.3. Enzymes predominantly used in biotechnology. From Rowe 2006. (Author's own table.)

Class	Function
oxidoreductases	oxidation-reduction involving transfer of electrons
transferases	transfer of certain functional groups such as: acyl, glycosyl (sugars), phosphate, methyl and sulfur groups
hydrolases	hydrolysis reactions (addition of water) with substrates such as amides, esters, epoxides, glycosides, peptides; esterification reactions
lyases	addition of groups to double bonds or formation of double bonds by removal of groups
isomerases	transfer of groups within molecules to yield isomeric forms; isomerization, racemization
ligases	formation of bonds by condensation reactions coupled to co-factor ATP or other nucleoside triphosphate cleavage
phytases	inactivation of phytic acids which bind mineral ions

depending on factors such as the type of EO used, the food matrix and the specific microorganisms present. Therefore, the concentration of EOs, as well as the application method, must be carefully considered to achieve the desired preservation effect without negatively impacting the sensory qualities of the food. Additionally, adherence to regulatory guidelines and safety considerations is crucial when using EOs in food products.

7.2.4   Distillation/solvent extraction

Traditionally used herbal substances include EOs and extracts, oleoresins, tinctures, distillates and juice concentrates (Arctander, 1960). Buds, flowers, fruits, peels, seeds, kernels, stems, leaves, barks and roots are used to produce these extracts. Extraction methods (see also Chapter 3) include expression, distillation and solvent extraction (including supercritical and subcritical fluids). While the (water vapour) volatile components are extracted during distillation, both volatile and non-volatile components are extracted during expression and solvent extraction, depending on the polarity of the solvents used.

As an agricultural by-product, fruit peels are a cost-effective source of natural flavourings for the food, pharmaceutical and cosmetics industries. The extraction of volatile flavour compounds such as EOs from various fruit peels using different modern extraction methods has only recently been evaluated (Liang *et al.*, 2020).

7.2.5   Spray-drying

Today, spray-drying is generally used in the pharmaceutical and food industries for the microencapsulation of EOs, as this process leads to particularly gentle drying due to the short contact time of the drying gas with the product (Mohammed 2020). In this context the flavouring (essential oil) is atomized into small droplets, which are then dried quickly to form a fine powder. Here, it is crucial for the respective product quality that the emulsion formed with the EO remains intact until the mixture is atomized. This is usually achieved by sufficient homogenization and the use of carrier materials that optimally support the stability of the emulsion (e.g. gum Arabic, maltodextrin) and at the same time serve as a carrier and encapsulation material for the flavour to be produced. The solids content of the starting material is usually maximized at a concentration and viscosity that enables successful atomization in order to use the drying equipment as efficiently as possible.

7.3   Health Aspects

To qualify as an aromatic and spice plant, a plant must contain phytochemicals that can be detected by our sensory system. Humans have olfactory receptors on the sensory cells in the nasopharyngeal region. The receptors (more than 350 different ones in humans) are membrane proteins (G-protein coupled receptor; Malnic *et al.*, 2004; Antunes and Simoes de Souza, 2016). Taste receptors are mainly localized on our tongue and can recognize compounds which taste salty, acidic, sweet, bitter and umami (Dutta Banik and Medler, 2022). More than 1000 genes have been detected which code for olfactory receptors and about 50 for taste receptors (Bachmanov *et al.*, 2014). Their expression differs between individuals and within the lifetime of a person.

7.3.1   Modes of action

Lipophilic phytochemicals, such as EOs, modulate the permeability and fluidity of biomembranes which surround the cells of all living organisms (Wink, 2015, 2022). The effect is dose dependent, and higher concentration can lyse biomembranes in animal cells or in bacteria, fungi and enveloped viruses. This effect is not specific, but was apparently useful in their million years of evolutionary history to protect themselves against herbivores and microbes (Wink, 2015).

Case in point: The task of the calcium channels.

It could be shown experimentally and in clinical studies that menthol has anti-spasmodic and carminative effects by inhibiting the calcium channels (Scarpellini *et al.*, 2023). Pungent compounds such as capsaicin or isothiocyanates bind to TRPV1 receptors (Tominaga and Tominaga, 2005; Leijon *et al.*, 2019; Shuba, 2020), which are calcium channels. Menthol and other monoterpenes which induce the feeling of cold affect another calcium channel, the temperature receptor (TRPM8) (Yin and Lee, 2020).

A number of proteins reside in biomembranes of human cells, including transporters, receptors and ion channels. In our context, calcium channels are of special importance as they can relax smooth muscle cells in the gut. When menthol and other lipophilic monoterpenes accumulate in biomembranes in the vicinity of calcium channels, they can inhibit their activity by altering the interaction of phospholipids with the channel proteins.

Bitter receptors mostly likely evolved in animals to detect and avoid toxic food (Jalševac *et al.*, 2022); this strategy works as most toxins exert a bitter taste in humans and other mammals. About 25 genes are known in humans that encode bitter receptors (TAS2R) (Lang *et al.*, 2023). Persons with a gene defect at the receptor gene TAS2R16 cannot taste bitter compounds, such as phenylthiocarbamide (PTC) or cyanide (Bachmanov *et al.*, 2014; Smail, 2019).

7.3.2   Medicinal properties of aromatic plants and spices

7.3.2.1   Antimicrobial properties

Disturbance of biomembranes (increase of membrane fluidity, pore formation) inhibits the growth of microorganisms or kills them directly. In traditional medicine and in phytotherapy, plants with lipophilic terpenoids, phenylpropanoids and saponins are employed in the treatment of microbial infections, respiratory conditions and skin infections (van Wyk and Wink, 2015; Wink, 2015, 2022).

These properties were probably one of the reasons why humans have selected aromatic plants and spices in food and beverages (Tables 7.1 and 7.2), because in the times before we had refrigerators, our food was contaminated by microbes (Wink, 2022). Cooking and the addition of spices could reduce the microbial contamination. Furthermore, aromatic and spicy food additives helped to improve the taste of food or beverages with a bad taste.

7.3.2.2   Antioxidant properties

Reactive oxygen species (ROS) can oxidize proteins, biomembranes and DNA in our cells. The oxidation of the DNA base guanosine leads to the formation of 8-oxoguanosine (8-oxoG). 8-oxoG no longer pairs with C (as guanosine would do) but with adenosine. This can lead to mutations and health conditions caused by genetic changes, such as cancer and many others (Wink, 2015, 2022; Roxo and Wink, 2022).

A number of antioxidant compounds are known from plants which can inactivate ROS, such as phenolics, carotenoids, ascorbic acid, allicin, isothiocyanates and even chlorophyll (Wink, 2022). Many spices contain phenolic compounds, allicin and isothiocyanates (Tables 7.1 and 7.2); all of them have substantial antioxidant properties. This could be another reason why spices are used in our nutrition (Wink, 2022).

7.3.2.3 Anti-inflammatory properties

Inflammation is a common symptom of many diseases and a response of the body to overcome them. Since the symptoms of inflammation (pain, fever) are not welcomed, we use anti-inflammatory drugs, such as cortisol or non-steroidal anti-inflammatory drugs (NSAID). A number of medicinal plants, spices (Tables 7.1 and 7.2) and phytoconstituents exhibit anti-inflammatory properties, including polyphenols, iridoid glucosides, salicylic acid and derivatives, and a number of terpenoids (chamazulene, sesquiterpene lactones) (van Wyk and Wink, 2015, 2017; Wink, 2015; Wink, 2022). The natural products can inhibit cyclooxygenases, phospholipases (PLA) or alkylate NFkB (allicin, helenalin) and other proteins of the inflammation pathway (Wink, 2015, 2022).

An important side effect of spices and aromatic plants is their anti-inflammatory properties, as they help to modulate acute and chronic inflammatory processes in our body.

7.3.2.4 Synergistic effects

Extracts from medicinal plants, EOs or spices are multi-component mixtures containing active secondary metabolites which can affect different molecular targets (Wink, 2008; Bunse *et al.*, 2022). Furthermore, the combination of individual components can result in an additive and, more interestingly, a synergistic effect.

The combination of two or three different terpenoids or alkaloids was demonstrated to exhibit a synergistic antimicrobial effect (Mulyaningsih *et al.*, 2010; Al-Ani *et al.*, 2015). Synergistic effect could be one reason why natural complex substances are still in use in medicine and food technology.

7.4 Aspects of Food Law

While EOs offer natural preservation benefits, their effectiveness can vary depending on factors such as the type of EO used, the food matrix and the specific microorganisms present. Furthermore, the concentration of EOs, as well as the application method, must be carefully considered to achieve the desired preservation effect without negatively impacting the sensory quality of the food.

Food laws and regulations related to the labelling of flavouring extracts, particularly when EOs are used for flavouring, can vary by country or region. However, some general aspects that are commonly addressed in food regulations are as follows:

- ingredient declaration;
- allergen information;
- natural/artificial flavours;
- adulterations;

- additives; and
- preservatives.

Although the FDA has attempted to define the term 'natural' in relation to food in general, the term 'natural flavouring' as used in the Code of Federal Regulations (CFR) refers to EO, oleoresin, essence or extract, protein hydrolysate and distillate. In addition, any product obtained by roasting, heating or enzymolysis that contains flavouring substances derived from spices, fruits or fruit juices, vegetable or vegetable juices, edible yeasts, herbs, barks, roots, leaves, including their fermentation products, and whose essential function in food is flavour rather than nutritional value, is classified as a natural flavouring.

> **Case in point: Generally Recognized As Safe (GRAS).**
>
> The Flavour and Essence Manufacturers Association (FEMA) has drawn up a so-called white list of approved flavouring chemicals and related substances. This list contains more than 4000 chemical compounds which are 'Generally Recognized As Safe (GRAS)', i.e. they may be used in certain quantities in food and do not have to be declared as individual substances on the food.

Another aspect of the regulation of flavourings concerns compliance with various dietary requirements. These can be based on faith (kosher, halal), health (allergen-free) or ethics (vegetarian, vegan, GMO-free and 'organic' labelling). The most widespread are the kosher regulations, where rabbinical committees have been dealing with the interactions between Mosaic law and the modern flavouring and chemical industry for many years (Salzer and Jones, 1998).

7.4.1 Essential oils as an important factor for securing employment in rural areas

The cultivation, harvesting and processing of plant-based raw materials have a significant impact on the employment situation of the rural population worldwide, particularly in less developed countries. Experts from IFEAT (International Federation of Essential Oils and Aroma Trades) estimate that over 10 million farmers and harvesters are saved from poverty thanks to aromatic plants, which are often produced on their small plots of land, sometimes using very old techniques (Greenhalgh, 2023).

Bibliography

ABDA (2024) Bundesvereinigung Deutscher Apothekerverbände (früher) Deutscher Arzneimittel-Codex, Neues Rezeptur-Formularium (DAC/NRF). Mediengruppe Deutscher Apotheker GmbH; Deutscher Apotheker Verlag; Govi-Verl. Pharmazeutischer Verlag, Eschborn, Stuttgart, Germany.

Abdel Rahman, A.N., Mohamed, A.A.-R., Mohammed, H.H., Elseddawy, N.M., Salem, G.A. et al. (2020) The ameliorative role of geranium (*Pelargonium graveolens*) essential oil against hepato-renal toxicity, immunosuppression, and oxidative stress of profenofos in common carp, *Cyprinus carpio* (L). *Aquaculture* 517, 734777.

Abelan, U.S., de Oliveira, A.C., Cacoci, É.S.P., Martins, T.E.A., Giacon, V.M. et al. (2022) Potential use of essential oils in cosmetic and dermatological hair products: A review. *Journal of Cosmetic Dermatology* 21, 1407–1418.

Abukhalil, M.H., Hussein, O.E., Aladaileh, S.H., Althunibat, O.Y., Al-Amarat, W. et al. (2021) Visnagin prevents isoproterenol-induced myocardial injury by attenuating oxidative stress and inflammation and upregulating Nrf2 signaling in rats. *Journal of Biochemical and Molecular Toxicology* 35(11), e22906.

Aburjai, T. and Natsheh, F.M. (2003) Plants used in cosmetics. *Phytotherapy Research: PTR* 17, 987–1000.

Agarwal, V. (2018) Complementary and alternative medicine provider knowledge discourse on holistic health. *Frontiers in Communication* 3.

Ahmad, A., Elisha, I.L., van Vuuren, S. and Viljoen, A. (2021) Volatile phenolics: A comprehensive review of the anti-infective properties of an important class of essential oil constituents. *Phytochemistry* 190, 112864.

Aichberger, L., Grafschalter, M., Fritsch, F., Gansinger, D., Hagmüller, W. et al. (2012) *Kräuter für Nutz- und Heimtier: Ratgeber für die Anwendung ausgewählter Heil- und Gewürzpflanzen*. Self-published.

Al-Ani, I., Zimmermann, S., Reichling, J. and Wink, M. (2015) Pharmacological synergism of bee venom and melittin with antibiotics and plant secondary metabolites against multi-drug resistant microbial pathogens. *Phytomedicine: International Journal of Phytotherapy and Phytopharmacology* 22, 245–255.

Albuquerque, K.R.S., Purgato, G.A., Píccolo, M.S., Rodrigues, F.F., Pizziolo, V.R. et al. (2023) Formulations of essential oils obtained from plants traditionally used as condiments or traditional medicine active against *Staphylococcus aureus* isolated from dairy cows with mastitis. *Letters in Applied Microbiology* 76(3).

Allenspach, M., Valder, C., Flamm, D., Grisoni, F. and Steuer, C. (2020) Verification of chromatographic profile of primary essential oil of *Pinus sylvestris* L. combined with chemometric analysis. *Molecules (Basel, Switzerland)* 25, 2973.

Alraddadi, B.G. and Shin, H.-J. (2022) Biochemical properties and cosmetic uses of *Commiphora myrrha* and *Boswellia serrata*. *Cosmetics* 9(6), 119.

Altay, Ö., Köprüalan, Ö., İlter, I., Koç, M., Ertekin, F.K. *et al.* (2024) Spray drying encapsulation of essential oils: Process efficiency, formulation strategies, and applications. *Critical Reviews in Food Science and Nutrition* 64(4), 1139–1157.

Álvarez-Martínez, F.J., Barrajón-Catalán, E., Herranz-López, M. and Micol, V. (2021) Antibacterial plant compounds, extracts and essential oils: An updated review on their effects and putative mechanisms of action. *Phytomedicine: International Journal of Phytotherapy and Phytopharmacology* 90, 153626.

Amat, S., Baines, D., Timsit, E., Hallewell, J. and Alexander, T.W. (2019) Essential oils inhibit the bovine respiratory pathogens *Mannheimia haemolytica, Pasteurella multocida* and *Histophilus somni* and have limited effects on commensal bacteria and turbinate cells *in vitro*. *Journal of Applied Microbiology* 126, 1668–1682.

American Botanical Council (ed.) (n.d.) Overview of Essential Oil Adulteration. Available at: https://www.herbalgram.org/search?search=adulteration%20essential%20oil (accessed 8 November 2024).

Ande, S.N. and Bakal, R.L. (2022) Potential herbal essential oils: Are they super natural skin protector? *Innovations in Pharmaceuticals and Pharmacotherapy* 2, 19–24. Available at: http://www.innpharmacotherapy.com/VolumeArticles/FullTextPDF/10235_04_IPP_10-AJ-2022-19.pdf (accessed 9 April 2025).

Angioni, A., Barra, A., Coroneo, V., Dessi, S. and Cabras, P. (2006) Chemical composition, seasonal variability, and antifungal activity of *Lavandula stoechas* L. ssp. *stoechas* essential oils from stem/leaves and flowers. *Journal of Agricultural and Food Chemistry* 54, 4364–4370.

Antunes, G. and Simoes de Souza, F.M. (2016) Olfactory receptor signaling. *Methods in Cell Biology* 132, 127–145.

Anwar, Y. and Siringoringo, V.S. (2020) Fractionation of citronella oil and identification of compounds by gas chromatography-mass spectrometry. *Pharmaceutical Sciences and Research* 7(3), 138–144.

Api, A.M., Basketter, D.A., Cadby, P.A., Cano, M.-F., Ellis, G. *et al.* (2008) Dermal sensitization quantitative risk assessment (QRA) for fragrance ingredients. *Regulatory Toxicology and Pharmacology* 52, 3–23.

Api, A.M., Basketter, D., Bridges, J., Cadby, P., Ellis, G. *et al.* (2020) Updating exposure assessment for skin sensitization quantitative risk assessment for fragrance materials. *Regulatory Toxicology and Pharmacology* 118, 104805.

Arctander, S. (1960) *Perfume and Flavor Materials of Natural Origin*. Orchard Innovations, Harlow, United Kingdom.

Arimura, G., Kost, C. and Boland, W. (2005) Herbivore-induced, indirect plant defences. *Biochimica et Biophysica Acta* 1734(2), 91–111.

Armstrong, J.S. (2006) Mitochondrial membrane permeabilization: The sine qua non for cell death. *BioEssays* 28(3), 253–260.

Arraiza, M.P. (2017) *Medicinal and aromatic plants: The basics of industrial application*. Bentham Science Publishers, Sharjah Airport International Free Trade Zone, UAE.

Arthur, C.L. and Pawliszyn, J. (1990) Solid phase microextraction with thermal desorption using fused silica optical fibers. *Analytical Chemistry* 62, 2145–2148.

Asadollahi-Baboli, M. and Aghakhani, A. (2015) Headspace adsorptive microextraction analysis of oregano fragrance using polyaniline-nylon-6 nanocomposite, GC-MS, and multivariate curve resolution. *International Journal of Food Properties* 18(7), 1613–1623.

Association of Official Agricultural Chemists (2023) *Official Methods of Analysis of AOAC INTERNATIONAL (OMA)*. Available at: https://www.aoac.org/official-methods-of-analysis (accessed 11 November 2024).

Atlantik Technological University (2022) ATU Marine Scientist's Pioneering Research on Reducing Microplastic Pollution Recognised in Europe. Available at: https://www.atu.ie/news/atu-marine-scientists-pioneering-research-on-reducing-microplastic-pollution-recognised-in-europe (accessed 10 December 2024).

Averbeck, D. and Averbeck, S. (1998) DNA photodamage, repair, gene induction and genotoxicity following exposures to 254 nm UV and 8-methoxypsoralen plus UVA in a eukaryotic cell system. *Photochemistry and Photobiology* 68(3), 289–295.

Averbeck, D., Averbeck, S., Dubertret, L., Young, A.R. and Morlière, P. (1990) Genotoxicity of bergapten and bergamot oil in *Saccharomyces cerevisiae*. *Journal of Photochemistry and Photobiology B: Biology* 7(2–4), 209–229.

Avila Gandra, E., Radünz, M., Helbig, E., Dellinghausen Borges, C. and Kuka Valente Gandra, T. (2018) A mini-review on encapsulation of essential oils. *Journal of Analytical & Pharmaceutical Research* 7(1).

Azevedo, V.M., Carvalho, R.A., Borges, S.V., Claro, P.I.C., Hasegawa, F.K. *et al.* (2019) Thermoplastic starch/whey protein isolate/rosemary essential oil nanocomposites obtained by extrusion process: Antioxidant polymers. *Journal of Applied Polymer Science* 136(23), 47619.

Bachmanov, A.A., Bosak, N.P., Lin, C., Matsumoto, I., Ohmoto, M. *et al.* (2014) Genetics of taste receptors. *Current Pharmaceutical Design* 20, 2669–2683.

Bakkali, F., Averbeck, S., Averbeck, D., Zhiri, A. and Idaomar, M. (2005) Cytotoxicity and gene induction by some essential oils in the yeast *Saccharomyces cerevisiae*. *Mutation Research* 585(1–2), 1–13.

Bakkali, F., Averbeck, S., Averbeck, D. and Idaomar, M. (2008) Biological effects of essential oils – a review. *Food and Chemical Toxicology* 46, 446–475.

Baldim, I., Paziani, M.H., Grizante Barião, P.H., Kress, M. and Oliveira, W.P. (2022) Nanostructured lipid carriers loaded with *Lippia sidoides* essential oil as a strategy to combat the multidrug-resistant *Candida auris*. *Pharmaceutics* 14(1), 180.

Banovac, D. (2012) Antimikrobielle Wirkung ausgewählter flüchtiger Verbindungen und ätherischer Öle auf luftgetragene Keime. MPharm thesis, University of Vienna, Austria.

Bañuelos-Hernández, A.E., Azadniya, E., Ramírez Moreno, E. and Morlock, G.E. (2020) Bioprofiling of Mexican *Plectranthus amboinicus* (Lour.) essential oil via planar chromatography–effect-directed analysis combined with direct analysis in real time high-resolution mass spectrometry. *Journal of Liquid Chromatography & Related Technologies* 43, 344–350.

Barbaud, A., Kurihara, F., Raison-Peyron, N., Milpied, B., Valois, A. *et al.* (2023) Allergic contact dermatitis from essential oil in consumer products: Mode of uses and value of patch tests with an essential oil series. Results of a French Study of the DAG (Dermato-Allergology group of the French Society of Dermatology). *Contact Dermatitis* 89(3), 190–197.

Barbieri, C. and Borsotto, P. (2018) Essential oils: Market and legislation. In: El-Shemy, H.A. (ed.) *Potential of Essential Oils*. Intech Open.

Barnes, J. (2007) *Herbal Medicines*, 3rd edn. Pharmaceutical Press, London and Chicago, Illinois.

Battisti, M.A., Constantino, L., Argenta, D.F., Reginatto, F.H., Pizzol, F.D. *et al.* (2024) Nanoemulsions and nanocapsules loaded with *Melaleuca alternifolia* essential oil for sepsis treatment. *Drug Delivery and Translational Research* 14(5), 1239–1252.

Beier, C., Demleitner, M., Hamm, D. and Danner, H. (2022) *Aromapraxis Heute: Ätherische Öle - Wirkung - Anwendung*. Elsevier, Munich.

Beigi, M., Torki-Harchegani, M. and Ghasemi Pirbalouti, A. (2018) Quantity and chemical composition of essential oil of peppermint (*Mentha × piperita* L.) leaves under different drying methods. *International Journal of Food Properties* 21(1), 267–276.

Bello, N.S., Antunes, F.T.T. and Gerhardt Martins, M. (2022) Essential oils as preservatives in cosmetics: An integrative review. *Open Access Journal of Biomedical Science* 4, 1511–1516. Available at: https://web.archive.org/web/20220203142652id_/https://biomedscis.com/pdf/OAJBS.ID.000387.pdf (accessed 13 November 2024).

Bento, M.H., Ouwehand, A.C., Tiihonen, K., Lahtinen, S., Nurminen, P. *et al.* (2013) Essential oils and their use in animal feeds for monogastric animals – effects on feed quality, gut microbiota, growth performance and food safety: A review. *Veterinární Medicína* 58(9), 449–458.

Bertea, C.M., Azzolin, C.M.M., Bossi, S., Doglia, G. and Maffei, M.E. (2005) Identification of an EcoRi restriction site for a rapid and precise determination of beta-asarone-free Acorus calamus cytotypes. *Phytochemistry* 66(5), 507–514.

Bhalla, Y., Gupta, V.K. and Jaitak, V. (2013) Anticancer activity of essential oils: A review. *Journal of the Science of Food and Agriculture* 93(15), 3643–3653.

Bhavaniramya, S., Vishnupriya, S., Al-Aboody, M.S., Vijayakumar, R. and Baskaran, D. (2019) Role of essential oils in food safety: Antimicrobial and antioxidant applications. *Grain & Oil Science and Technology* 2(2), 49–55.

Bialas, I., Zelent-Kraciuk, S. and Jurowski, K. (2023) The skin sensitisation of cosmetic ingredients: Review of actual regulatory status. *Toxics* 11(4), 392.

Bismarck, D., Schneider, M. and Müller, E. (2017) Antibakterielle *In-vitro*-Wirksamkeit ätherischer Öle gegen veterinärmedizinisch relevante Keime klinischer Isolate von Hunden, Katzen und Pferden. *Complementary Medicine Research* 24(3), 153–163.

Bitterling, H., Lorenz, P., Vetter, W., Conrad, J., Kammerer, D.R. *et al.* (2020) Rapid spectrophotometric method for assessing hydroperoxide formation from terpenes in essential oils upon oxidative conditions. *Journal of Agricultural and Food Chemistry* 68, 9576–9584.

Bitterling, H., Lorenz, P., Vetter, W., Kammerer, D.R. and Stintzing, F.C. (2022a) Photo-protective effects of selected furocoumarins on β-pinene, R-(+)-limonene and γ-terpinene upon UV-A irradiation. *Journal of Photochemistry and Photobiology A: Chemistry* 424, 113623.

Bitterling, H., Mailänder, L., Vetter, W., Kammerer, D.R. and Stintzing, F.C. (2022b) Photo-protective effects of furocoumarins on terpenes in lime, lemon and bergamot essential oils upon UV light irradiation. *European Food Research and Technology* 248, 1049–1057.

Blerot, B., Martinelli, L., Prunier, C., Saint-Marcoux, D., Legrand, S. *et al.* (2018) Functional analysis of four terpene synthases in rose-scented *Pelargonium* cultivars (*Pelargonium* × *hybridum*) and evolution of scent in the *Pelargonium* genus. *Frontiers in Plant Science* 9, 1435.

Blowey, R.W. and Edmondson, P. (2010) *Mastitis Control in Dairy Herds*, 2nd edn. CAB International, Wallingford, UK.

Bounaas, K., Bouzidi, N., Daghbouche, Y., Garrigues, S., Guardia, M. de *et al.* (2018) Essential oil counterfeit identification through middle infrared spectroscopy. *Microchemical Journal* 139, 347–356.

Brain, K.R., Green, D.M., Jones, A.C., Walters, K.A., Api, A.M. *et al.* (2022) In vitro human skin absorption of linalool: Effects of vehicle composition, evaporation and occlusion on permeation and distribution. *International Journal of Pharmaceutics* 622, 121826.

Brendieck-Worm, C. and Melzig, M.F. (eds) (2021) *Phytotherapie in der Tiermedizin*, 2nd edn. Georg Thieme Verlag, Stuttgart, Germany and New York.

Brendieck-Worm, C., Klarer, F. and Stöger, E. (2021) *Heilende Kräuter für Tiere: Pflanzliche Hausmittel für Heim- und Nutztiere*, 3rd edn. Haupt Verlag, Bern.

Brynzak-Schreiber, E., Schögl, E., Bapp, C., Cseh, K., Kopatz, V. *et al.* (2024) Microplastics role in cell migration and distribution during cancer cell division. *Chemosphere* 353, 141463.

Bundesinstitut für Arzneimittel und Medizinprodukte (2023) *Europäisches Arzneibuch [European Pharmacopoeia, PhEur]*, 11th edn. Deutscher Apotheker Verlag, Stuttgart, Germany.

Bundesinstitut für Risikobewertung, BfR (2003) Opinion of the Federal Institute for Risk Assessment (BfR) towards the use of undiluted tea-tree oil as a cosmetic, 1st September 2003. Available at: https://www.bfr.bund.de/cm/349/use_of_undiluted_tea_tree_oil_as_a_cosmetic.pdf (accessed 13 November 2024).

Bundesinstitut für Risikobewertung, BfR (2008) Frequently Asked Questions (FAQs) about the use of essential oils, 28th February 2008. Available at: https://www.bfr.bund.de/cm/343/fragen_und_antworten_zur_anwendung_von_aetherischen_oelen.pdf (accessed 13 November 2024).

Bundesinstitut für Risikobewertung, BfR (2014) Determination of Pyrrolizidine Alkaloids (PA) in Plant Material by SPE-LC-MS/MS, Method Protocol BfR-PA-Tea-2.0/2014. Available at: https://www.bfr.bund.de/cm/349/determination-of-pyrrolizidine-alkaloids-pa-in-plant-material.pdf (accessed 11 November 2024).

Bunse, M., Daniels, R., Gründemann, C., Heilmann, J., Kammerer, D.R. *et al.* (2022) Essential oils as multicomponent mixtures and their potential for human health and well-being. *Frontiers in Pharmacology* 13, 956541.

Buriani, A., Fortinguerra, S., Sorrenti, V., Caudullo, G. and Carrara, M. (2020) Essential oil phytocomplex activity, a review with a focus on multivariate analysis for a network pharmacology-informed phytogenomic approach. *Molecules (Basel, Switzerland)* 25, 1833.

Cadby, P.A., Troy, W.R. and Vey, M.G.H. (2002) Consumer exposure to fragrance ingredients: Providing estimates for safety evaluation. *Regulatory Toxicology and Pharmacology* 36(3), 246–252.

Cadé, D., Cole, E.T., Mayer, J.P. and Wittwer, F. (1986) Liquid filled and sealed hard gelatin capsules. *Drug Development and Industrial Pharmacy* 12, 2289–2300.

Caissard, J.-C., Joly, C., Bergougnoux, V., Hugueney, P., Mauriat, M. *et al.* (2004) Secretion mechanisms of volatile organic compounds in specialized cells of aromatic plants. *Recent Research Developments in Cell Biology* 2, 1–15. Available at: https://hal.archives-ouvertes.fr/ujm-00081423/

Camele, I., Elshafie, H.S., Caputo, L. and De Feo, V. (2019) Anti-quorum sensing and antimicrobial effect of mediterranean plant essential oils against phytopathogenic bacteria. *Frontiers in Microbiology* 10, 2619.

Canibe, N., Højberg, O., Kongsted, H., Vodolazska, D., Lauridsen, C. *et al.* (2022) Review on preventive measures to reduce post-weaning diarrhoea in piglets. *Animals* 12(9), 2585.

Carson, C.F., Mee, B.J. and Riley, T.V. (2002) Mechanism of action of *Melaleuca alternifolia* (tea tree) oil on *Staphylococcus aureus* determined by time-kill, lysis, leakage, and salt tolerance assays and electron microscopy. *Antimicrobial Agents and Chemotherapy* 46(6), 1914–1920.

Carson, C.F., Hammer, K.A. and Riley, T.V. (2006) *Melaleuca alternifolia* (Tea Tree) oil: A review of antimicrobial and other medicinal properties. *Clinical Microbiology Reviews* 19(1), 50–62.

Carvalho, I.T., Estevinho, B.N. and Santos, L. (2016) Application of microencapsulated essential oils in cosmetic and personal healthcare products - a review. *International Journal of Cosmetic Science* 38(2), 109–119.

Castelani, L., Arcaro, J.R.P., Braga, J.E.P., Bosso, A.S., Moura, Q. *et al.* (2019) Short communication: Activity of nisin, lipid bilayer fragments and cationic nisin-lipid nanoparticles against multidrug-resistant *Staphylococcus* spp. isolated from bovine mastitis. *Journal of Dairy Science* 102, 678–683.

Cavanagh, H.M.A. and Wilkinson, J.M. (2002) Biological activities of lavender essential oil. *Phytotherapy Research: PTR* 16, 301–308.

CBI (2024) What is the demand for natural ingredients for cosmetics on the European market? Available at: https://www.cbi.eu/sites/default/files/pdf/research/1137.pdf (accessed 13 November 2024).

Cebi, N., Taylan, O., Abusurrah, M. and Sagdic, O. (2020) Detection of orange essential oil, isopropyl myristate, and benzyl alcohol in lemon essential oil by FTIR spectroscopy combined with chemometrics. *Foods (Basel, Switzerland)* 10(1), 27.

Celebi Sozener, Z., Özbey Yücel, Ü., Altiner, S., Ozdel Oztürk, B., Cerci, P. *et al.* (2022a) The external exposome and allergies: From the perspective of the epithelial barrier hypothesis. *Frontiers in Allergy* 3, 887672.

Celebi Sozener, Z., Ozdel Ozturk, B., Cerci, P., Turk, M., Gorgulu Akin, B. *et al.* (2022b) Epithelial barrier hypothesis: effect of the external exposome on the microbiome and epithelial barriers in allergic disease. *Allergy* 77, 1418–1449.

Cengiz Çallıoğlu, F. and Kesici Güler, H. (2020) Production of essential oil-based composite nanofibers by emulsion electrospinning. *Pamukkale University Journal of Engineering Sciences* 26(7), 1178–1185.

Charles Dorni, A.I., Amalraj, A., Gopi, S., Varma, K. and Anjana, S.N. (2017) Novel cosmeceuticals from plants—an industry guided review. *Journal of Applied Research on Medicinal and Aromatic Plants* 7, 1–26.

Chauiyakh, O., El Fahime, E., Aarabi, S., Ninich, O., Bentata, F. *et al.* (2023) A systematic review on chemical composition and biological activities of cedar oils and extracts. *Research Journal of Pharmacy and Technology* 16, 3875–3883.

Chen, L., Liu, Y., Xu, D., Zhang, N., Chen, Y. *et al.* (2024) Beta-myrcene as a sedative-hypnotic component from lavender essential oil in DL-4-chlorophenylalanine-induced-insomnia mice. *Pharmaceuticals (Basel, Switzerland)* 17, 1161.

Choi, S.Y., Kang, P., Lee, H.S. and Seol, G.H. (2014) Effects of inhalation of essential oil of *Citrus aurantium* L. var. amara on menopausal symptoms, stress, and estrogen in postmenopausal women: A randomized controlled trial. *Evidence-Based Complementary and Alternative Medicine* 2014, 796518.

Chou, T.-C. (2006) Theoretical basis, experimental design, and computerized simulation of synergism and antagonism in drug combination studies. *Pharmacological Reviews* 58(3), 621–681.

Christoffers, W.A., Blömeke, B., Coenraads, P.-J. and Schuttelaar, M.-L.A. (2014) The optimal patch test concentration for ascaridole as a sensitizing component of tea tree oil. *Contact Dermatitis* 71(3), 129–137.

Cimino, C., Maurel, O.M., Musumeci, T., Bonaccorso, A., Drago, F. *et al.* (2021) Essential oils: Pharmaceutical applications and encapsulation strategies into lipid-based delivery systems. *Pharmaceutics* 13(3), 327.

Ciuman, R.R. (2012) Phytotherapeutic and naturopathic adjuvant therapies in otorhinolaryngology. *European Archives of Oto-Rhino-Laryngology* 269(2), 389–397.

Clark, D., Edwards, E., Murray, P. and Langevin, H. (2021) Implementation science methodologies for complementary and integrative health research. *Journal of Alternative and Complementary Medicine (New York, N.Y.)* 27(S1), S7–S13.

Clery, R.A., Armendi, A., Franco, V., Furrer, S., Genereux, J.C. *et al.* (2022) Chemical diversity of citrus leaf essential oils. *Chemistry & Biodiversity* 19(3), e202100963.

Cluzel, M., Hais, G., Irizar, A., Lenouvel, V., Nash, J.F. *et al.* (2022) Absence of phototoxicity/photoirritation potential of bergamottin determined *in vitro* using OECD TG 432. *Regulatory Toxicology and Pharmacology* 136, 105281.

Cohen, S.M., Eisenbrand, G., Fukushima, S., Gooderham, N.J., Guengerich, F.P. *et al.* (2019) FEMA GRAS assessment of natural flavor complexes: Citrus-derived flavoring ingredients. *Food and Chemical Toxicology* 124, 192–218.

Cosmetics Europe (2017) Consumer insights 2017. Available at: https://cosmeticseurope.eu/files/6114/9738/2777/CE_Consumer_Insights_2017.pdf (accessed 9 April 2025).

Cuchet, A., Anchisi, A., Schiets, F., Clément, Y., Lantéri, P. *et al.* (2021) Determination of enantiomeric and stable isotope ratio fingerprints of active secondary metabolites in neroli (*Citrus aurantium* L.) essential oils for authentication by multidimensional gas chromatography and GC-C/P-IRMS. *Journal of Chromatography B* 1185, 123003.

Cuesta, F., Paida, G. and Buele, I. (2020) Influence of olfactory and visual sensory stimuli in the perfume-purchase decision. *International Review of Management and Marketing* 10, 63–71.

Da Veiga, R.D.S., Da Aparecida Silva-Buzanello, R., Corso, M.P. and Canan, C. (2019) Essential oils microencapsulated obtained by spray drying: A review. *Journal of Essential Oil Research* 31(6), 457–473.

Daniels, R. and Mittermaier, E.M. (1995) Influence of pH adjustment on microcapsules obtained from complex coacervation of gelatin and acacia. *Journal of Microencapsulation* 12, 591–599.

Daning, D.A.R., Yusiati, L.M., Hanim, C. and Widyobroto, B.P. (2020) The use of essential oils as rumen modifier in dairy cows. *Indonesian Bulletin of Animal and Veterinary Sciences* 30(4), 189.

Darwish, A.A., Fawzy, M., Osman, W.A.-L. and El Ebissy, E.A. (2021) Clinicopathological and bacteriological studies on lamb bacterial enteritis and monitoring the oregano oil and vitamins A,D3,E effect on its treatment. *Journal of Advanced Veterinary and Animal Research* 8(2), 291–299.

Day, J. (2013) Botany meets archaeology: People and plants in the past. *Journal of Experimental Botany* 64, 5805–5816.

De Flora, S. and Ramel, C. (1988) Mechanisms of inhibitors of mutagenesis and carcinogenesis. Classification and overview. *Mutation Research* 202(2), 285–306.

de Groot, A.C. and Schmidt, E. (2017) Essential oils, part VI: Sandalwood oil, ylang-ylang oil, and jasmine absolute. *Dermatitis: Contact, Atopic, Occupational, Drug* 28(1), 14–21.

De Groot, A.C., Beverdam, E.G., Ayong, C.T., Coenraads, P.J. and Nater, J.P. (1988) The role of contact allergy in the spectrum of adverse effects caused by cosmetics and toiletries. *Contact Dermatitis* 19(3), 195–201.

de Vincenzi, M., Silano, M., Maialetti, F. and Scazzocchio, B. (2000) Constituents of aromatic plants: II. Estragole. *Fitoterapia* 71, 725–729.

Deigendesch, J. (1785) *Nachrichters Roß-Arzneybuch nebst einem Anhang von Rindvieh-Arzneyen*, 2nd edn. Saur, Frankfurt and Leipzig.

Deutscher Apotheker Verlag (2023) *Deutsches Arzneibuch 2023 [German Pharmacopoeia, DAB 2023]*, 12th edn. Deutscher Apotheker Verlag, Stuttgart, Germany.

Diepgen, T.L., Ofenloch, R.F., Bruze, M., Bertuccio, P., Cazzaniga, S. *et al.* (2016) Prevalence of contact allergy in the general population in different European regions. *The British Journal of Dermatology* 174(2), 319–329.

Dierings de Souza, E.J., Kringel, D.H., Guerra Dias, A.R. and Da Rosa Zavareze, E. (2021) Polysaccharides as wall material for the encapsulation of essential oils by electrospun technique. *Carbohydrate Polymers* 265, 118068.

Do, T.K.T., Hadji-Minaglou, F., Antoniotti, S. and Fernandez, X. (2015) Authenticity of essential oils. *TrAC Trends in Analytical Chemistry* 66, 146–157.

Dobreva, K. and Dimov, M.D. (2021) Study of the changes in the chemical composition of Bulgarian dill essential oils. *IOP Conference Series: Materials Science and Engineering* 1031.

Dolcini, J., Chiavarini, M., Firmani, G., Ponzio, E., D'Errico, M.M. *et al.* (2024) Consumption of bottled water and chronic diseases: A nationwide cross-sectional study. *International Journal of Environmental Research and Public Health* 21, 1074.

Dontje, A.E.W.K., Schuiling-Veninga, C.C.M., van Hunsel, F.P.A.M., Ekhart, C., Demirci, F. *et al.* (2024) The therapeutic potential of essential oils in managing inflammatory skin conditions: A scoping review. *Pharmaceuticals (Basel, Switzerland)* 17, 571.

Dubey, V.S., Bhalla, R. and Luthra, R. (2003) An overview of the non-mevalonate pathway for terpenoid biosynthesis in plants. *Journal of Biosciences* 28(5), 637–646.

Dudareva, N., Negre, F., Nagegowda, D.A. and Orlova, I. (2006) Plant volatiles: Recent advances and future perspectives. *Critical Reviews in Plant Sciences* 25(5), 417–440.

Dufault, R.J., Hassell, R., Rushing, J.W., McCutcheon, G., Shepard, M. *et al.* (2001) Revival of herbalism and its roots in medicine. *Journal of Agromedicine* 7(2), 21–29.

Duff, G.C. and Galyean, M.L. (2007) Board-invited review: Recent advances in management of highly stressed, newly received feedlot cattle. *Journal of Animal Science* 85(3), 823–840.

Dupuis, V., Cerbu, C., Witkowski, L., Potarniche, A.-V., Timar, M.C. *et al.* (2022) Nanodelivery of essential oils as efficient tools against antimicrobial resistance: A review of the type and physical-chemical properties of the delivery systems and applications. *Drug Delivery* 29(1), 1007–1024.

Dutta Banik, D. and Medler, K.F. (2022) Taste receptor signaling. *Handbook of Experimental Pharmacology* 275, 33–52.

Eamon, W. (2000) Alchemy in popular culture: Leonardo Fioravanti and the search for the philosopher's stone. *Early Science and Medicine* 5(2), 196–213.

Ebani, V.V. and Mancianti, F. (2020) Use of essential oils in veterinary medicine to combat bacterial and fungal infections. *Veterinary Sciences* 7(4), 193.

Ebani, V.V., Najar, B., Bertelloni, F., Pistelli, L., Mancianti, F. *et al.* (2018) Chemical composition and *in vitro* antimicrobial efficacy of sixteen essential oils against *Escherichia coli* and *Aspergillus fumigatus* isolated from poultry. *Veterinary Sciences* 5(3), 62.

Ebani, V.V., Nardoni, S., Bertelloni, F., Tosi, G., Massi, P. *et al.* (2019) In vitro antimicrobial activity of essential oils against *Salmonella enterica* serotypes enteritidis and Typhimurium strains isolated from poultry. *Molecules (Basel, Switzerland)* 24(900).

ECHA (2022) Essential Oils. Available at: https://echa.europa.eu/de/support/substance-identification/sector-specific-support-for-substance-identification/essential-oils (accessed 14 April 2022).

EDQM (2023) *Guide for the Elaboration of Monographs on Herbal Drugs and Herbal Drug Preparations: European Pharmacopoeia*. Available at: https://www.edqm.eu/en/d/1766460 (accessed 7 April 2025).

EDQM (2024) Guidance on Essential Oils in Cosmetic Products. Available at: https://www.edqm.eu/en/guidance-on-essential-oils-in-cosmetic-products (accessed 6 November 2024).

El Hachlafi, N., Benkhaira, N., Al-Mijalli, S.H., Mrabti, H.N., Abdnim, R. *et al.* (2023) Phytochemical analysis and evaluation of antimicrobial, antioxidant, and antidiabetic activities of essential oils from moroccan medicinal plants: *Mentha suaveolens, Lavandula stoechas,* and *Ammi visnaga. Biomedicine & Pharmacotherapy* 164, 114937.

European Commission (2002) Regulation (EC) No 178/2002 of the European Parliament and of the Council of 28 January 2002 laying down the general-principles and requirements of food law, establishing the European Food Safety Authority and laying down procedures in matters of food safety. Available at: https://eur-lex.europa.eu/legal-content/EN/TXT/PDF/?uri=CELEX:32002R0178 (accessed 11 November 2024).

European Commission (2005) Regulation (EC) No. 396/2005 of the European Parliament and of the Council of 23 February 2005 on maximum residue levels of pesticides in or on food and feed of plantand animal origin and amending Council Directive 91/414/EEC. Available at: https://eur-lex.europa.eu/legal-content/EN/TXT/PDF/?uri=CELEX:32005R0396 (accessed 8 November 2024).

European Commission (2006) Regulation (EC) No 1881/2006 of 19 December 2006 setting maximum levels for certain contaminants in foodstuffs. Available at: https://eur-lex.europa.eu/legal-content/EN/TXT/?uri=CELEX%3A32006R1881& qid=1743174094587 (accessed 8 November 2024).

European Commission (2009) Directive 2009/32/EC of the European Parliament and of the Council of 23 April 2009 on the approximation of the laws of the Member States on extraction solvents used in the production of foodstuffs and food ingredients. Available at: https://eur-lex.europa.eu/legal-content/EN/TXT/PDF/?uri=CELEX:02009L0032-20230216 (accessed 11 November 2024).

European Commission (2009) Regulation (EC) No 1223/2009 of the European Parliament and of the Council of 30 November 2009 on cosmetic products. Available at: https://eur-lex.europa.eu/eli/reg/2009/1223/oj (accessed 13 November 2024).

European Commission (2011) Regulation (EU) No 1169/2011 of the European Parliament and of the Council of 25 October 2011 on the provision of food information to consumers. Available at: http://www.data.europa.eu/eli/reg/2011/1169/oj (accessed 8 November 2024).

European Commission (2017) Regulation (EU) 2017/625 of the European Parliament and of the Council of 15 March 2017 on official controls and other official activities performed to ensure the application of food and feed law, rules on animal health and welfare, planthealth and plant protection products. Available at: https://eur-lex.europa.eu/legal-content/EN/TXT/?uri=CELEX:32017R0625 (accessed 11 November 2024).

European Commission (2022) Regulation (EU) No. 2022/1531 amending Regulation (EC) No 1223/2009 of the European Parliament and of the Council as regards the use in cosmetic products of certain substances classified as carcinogenic, mutagenic or toxic for reproduction and correcting that regulation. Available at: https://eur-lex.europa.eu/eli/reg/2022/1531/oj (accessed 13 November 2024).

European Commission (2023) Regulation (EU) 2023/1545 of 26 July 2023 amending Regulation (EC) No 1223/2009 of the European Parliament and of the Council as regards labelling of fragrance allergens in cosmetic products. Available at: https://eur-lex.europa.eu/eli/reg/2023/1545/oj (accessed 14 November 2024).

European Medicines Agency (n.d.) Herbal drugs monographs. Available at: https://www.ema.europa.eu/en/human-regulatory-overview/herbal-medicinal-products/european-union-monographs-list-entries (accessed 11 November 2024).

European Union (2016) EU Veterinary Medicinal Product Database - Data Europa EU. Available at: https://data.europa.eu/data/datasets/eu-veterinary-medicinal-product-database?locale=en (accessed 12 November 2024).

Evangelista, A.G., Corrêa, J.A.F., Pinto, A.C.S.M. and Luciano, F.B. (2022) The impact of essential oils on antibiotic use in animal production regarding antimicrobial resistance - a review. *Critical Reviews in Food Science and Nutrition* 62, 5267–5283.

Ferreira, T.S., Moreira, C.Z., Cária, N.Z., Victoriano, G., Silva Jr, W.F. *et al.* (2014) Phytotherapy: An introduction to its history, use and application. *Revista Brasileira de Plantas Medicinais* 16(2), 290–298.

Fischman, S.L., Aguirre, A. and Charles, C.H. (2004) Use of essential oil-containing mouthrinses by xerostomic individuals: Determination of potential for oral mucosal irritation. *American Journal of Dentistry* 17(1), 23–26.

Flemming, M., Kraus, B., Rascle, A., Jürgenliemk, G., Fuchs, S. *et al.* (2015) Revisited anti-inflammatory activity of matricine in vitro: Comparison with chamazulene. *Fitoterapia* 106, 122–128.

Fortineau, A.-D. (2004) Chemistry perfumes your daily life. *Journal of Chemical Education* 81, 45.

Francois-Newton, V., Brown, A., Andres, P., Mandary, M.B., Weyers, C. *et al.* (2021) Antioxidant and anti-aging potential of Indian sandalwood oil against environmental stressors *in vitro* and *ex vivo*. *Cosmetics* 8(2), 53.

Franzios, G., Mirotsou, M., Hatziapostolou, E., Kral, J., Scouras, Z.G. *Mavragani-Tsipidou*, P. (1997) Insecticidal and genotoxic activities of mint essential oils. *Journal of Agricultural and Food Chemistry* 45, 2690–2694.

Freitas, I.R. and Cattelan, M.G. (2018) Antimicrobial and antioxidant proper-ties of essential oils in food systems—an overview. In: Holban, A.M. and Grumezescu, A.M. (eds) *Handbook of Food Bioengineering, Volume 10: Microbial Contamination and Food Degradation*. Academic Press, Oxford, UK, pp. 443–470.

Frix, A. (2023) Essential Oils and Forest Extracts: Complex, Unique and Vulnerable Industries Facing EU Green Deal. Available at: https://www.perfumerflavorist.com/fragrance/regulatory-research/article/22863293/alain-frix-discusses-threat-to-essential-oils-industry-due-to-eu-green-deal (accessed 6 November 2024).

Fröhner, E. (1896) *Lehrbuch der Arzneimittellehre für Thierärzte*, 4th edn. Enke, Stuttgart, Germany.

Fu, H.-R., Li, X.-S., Zhang, Y.-H., Feng, B.-B. and Pan, L.-H. (2020) Visnagin ame-liorates myocardial ischemia/reperfusion injury through the promotion of autophagy and the inhibition of apoptosis. *European Journal of Histochemistry* 64(Suppl. 2).

Fürst, R. and Zündorf, I. (2015) Evidence-based phytotherapy in Europe: Where do we stand? *Planta Medica* 81(12–13), 962–967.

Geier, J., Schnuch, A., Lessmann, H. and Uter, W. (2015a) Reactivity to sorbitan sesquioleate affects reactivity to fragrance mix I. *Contact Dermatitis* 73(5), 296–304.

Geier, J., Uter, W., Lessmann, H. and Schnuch, A. (2015b) Fragrance mix I and II: Results of breakdown tests. *Flavour and Fragrance Journal* 30(4), 264–274.

Geier, J., Schubert, S., Reich, K., Skudlik, C., Ballmer-Weber, B. *et al.* (2022) Contact sensitization to essential oils: IVDK data of the years 2010-2019. *Contact Dermatitis* 87(1), 71–80.

Geppner, L., Grammatidis, S., Wilfing, H. and Henjakovic, M. (2024) First evidence of the possible influence of avoiding daily liquid intake from plastic and glass beverage bottles on blood pressure in healthy volunteers. *Microplastics* 3(3), 419–432.

Gharsallaoui, A., Roudaut, G., Chambin, O., Voilley, A. and Saurel, R. (2007) Applications of spray-drying in microencapsulation of food ingredients: An overview. *Food Research International* 40, 1107–1121.

Ghodrati, M., Farahpour, M.R. and Hamishehkar, H. (2019) Encapsulation of pep-permint essential oil in nanostructured lipid carriers: *In-vitro* antibacterial activ-ity and accelerative effect on infected wound healing. *Colloids and Surfaces A: Physicochemical and Engineering Aspects* 564, 161–169.

Ghosh, S., Sinha, J.K., Ghosh, S., Vashisth, K., Han, S. *et al.* (2023) Microplastics as an emerging threat to the global environment and human health. *Sustainability* 15(14), 10821.

Giannenas, I., Bonos, E., Christaki, E. and Florou-Paneri, P. (2013) Essential oils and their applications in animal nutrition. *Medicinal & Aromatic Plants* 2(6), 140.

Gionfriddo, F., Postorino, E. and Calabrò, G. (2004) Elimination furocoumarins in bergamot peel oil. *Perfumer & Flavorist* 29. Available at: https://img.perfum-erflavorist.com/files/base/allured/all/document/2016/02/pf.PF_29_05_048_04.pdf (accessed 9 April 2025).

Gomes-Carneiro, M.R., Felzenswalb, I. and Paumgartten, F.J. (1998) Mutagenicity testing (+/-)-camphor, 1,8-cineole, citral, citronellal, (-)-menthol and terpineol with the Salmonella/microsome assay. *Mutation Research* 416(1–2), 129–136.

Gomes-Carneiro, M.R., Dias, D.M., De-Oliveira, A.C. and Paumgartten, F.J. (2005) Evaluation of mutagenic and antimutagenic activities of alpha-bisabolol in the Salmonella/microsome assay. *Mutation Research* 585(1–2), 105–112.

Gonfa, Y.H., Tessema, F.B., Bachheti, A., Rai, N., Tadesse, M.G. *et al.* (2023) Anti-inflammatory activity of phytochemicals from medicinal plants and their nanoparticles: A review. *Current Research in Biotechnology* 6, 100152.

Góra, J. and Brud, W. (1983) Progress in synthesis of sensory important trace components of essential oils and natural flavours. *Die Nahrung* 27(5), 413–428.

Götz, M.E., Eisenreich, A., Frenzel, J., Sachse, B. and Schäfer, B. (2023) Occurrence of alkenylbenzenes in plants: Flavours and possibly toxic plant metabolites. *Plants (Basel, Switzerland)* 12(11), 2075.

Greenhalgh, P. (2023) *IFEAT Socio-Economic Report Orange*. Available at: https://ifeat.org/wp-content/uploads/2023/10/IFEATWORLD-Autumn-2023-Socio-Economic-Report-Orange.pdf (accessed 4 April 2025).

Guba, R. (2001) Toxicity myths - essential oils and their carcinogenic potential. *International Journal of Aromatherapy* 11(2), 76–83.

Guin, J.D., Meyer, B.N., Drake, R.D. and Haffley, P. (1984) The effect of quenching agents on contact urticaria caused by cinnamic aldehyde. *Journal of the American Academy of Dermatology* 10(1), 45–51.

Gullapalli, R.P. (2010) Soft gelatin capsules (softgels). *Journal of Pharmaceutical Sciences* 99, 4107–4148.

Gupta, S., Walia, A. and Malan, R. (2011) Phytochemistry and pharmacology of *Cedrus deodera*: An overview. *International Journal of Pharmaceutical Sciences and Research* 2, 2010–2020.

Guyader, S., Thomas, F., Jamin, E., Grand, M., Akoka, S. *et al.* (2019) Combination of 13 C and 2 H SNIF - NMR isotopic fingerprints of vanillin to control its precursors. *Flavour and Fragrance Journal* 34(2), 133–144.

Guzelmeric, E., Vovk, I. and Yesilada, E. (2015) Development and validation of an HPTLC method for apigenin 7-O-glucoside in chamomile flowers and its application for fingerprint discrimination of chamomile-like materials. *Journal of Pharmaceutical and Biomedical Analysis* 107, 108–118.

Guzmán, E. and Lucia, A. (2021) Essential oils and their individual components in cosmetic products. *Cosmetics* 8(4), 114.

Halla, N., Fernandes, I.P., Heleno, S.A., Costa, P., Boucherit-Otmani, Z. *et al.* (2018) Cosmetics preservation: A review on present strategies. *Molecules (Basel, Switzerland)* 23, 1571.

Hallier, A., Noirot, V., Medina, B., Leboeuf, L. and Cavret, S. (2013) Development of a method to determine essential oil residues in cow milk. *Journal of Dairy Science* 96, 1447–1454.

Hamer, H.M., Jonkers, D., Venema, K., Vanhoutvin, S., Troost, F.J. *et al.* (2008) Review article: The role of butyrate on colonic function. *Alimentary Pharmacology & Therapeutics* 27(2), 104–119.

Hammoud, Z., Gharib, R., Fourmentin, S., Elaissari, A. and Greige-Gerges, H. (2019) New findings on the incorporation of essential oil components into liposomes composed of lipoid S100 and cholesterol. *International Journal of Pharmaceutics* 561, 161–170.

Hans, M., Deeksha, Nayik, G.A. and Salaria, A. (2023) Clary sage essential oil. In: Nayik, G.A. and Ansari, M.J. (eds) *Essential Oils*. Elsevier, Amsterdam, pp. 459–478.

Hansen, T., Risborg, M.S. and Steen, C.D. (2012) Understanding consumer purchase of free-of cosmetics: A value-driven TRA approach. *Journal of Consumer Behaviour* 11(6), 477–486.

Hansen, A.-M.S., Fromberg, A. and Frandsen, H.L. (2014) Authenticity and traceability of vanilla flavors by analysis of stable isotopes of carbon and hydrogen. *Journal of Agricultural and Food Chemistry* 62, 10326–10331.

Hartman, P.E. and Shankel, D.M. (1990) Antimutagens and anticarcinogens: A survey of putative interceptor molecules. *Environmental and Molecular Mutagenesis* 15(3), 145–182.

Hasheminejad, G. and Caldwell, J. (1994) Genotoxicity of the alkenylbenzenes alpha- and beta-asarone, myristicin and elimicin as determined by the UDS assay in cultured rat hepatocytes. *Food and Chemical Toxicology* 32, 223–231.

Heghes, S.C., Vostinaru, O., Rus, L.M., Mogosan, C., Iuga, C.A. *et al.* (2019) Antispasmodic effect of essential oils and their constituents: A review. *Molecules (Basel, Switzerland)* 24, 1675.

Hemaiswarya, S., Kruthiventi, A.K. and Doble, M. (2008) Synergism between natural products and antibiotics against infectious diseases. *Phytomedicine: International Journal of Phytotherapy and Phytopharmacology* 15, 639–652.

Herman, A. (2014) Comparison of antimicrobial activity of essential oils, plant extracts and methylparaben in cosmetic emulsions: 2 months study. *Indian Journal of Microbiology* 54(3), 361–364.

Herman, A., Herman, A.P., Domagalska, B.W. and Młynarczyk, A. (2013) Essential oils and herbal extracts as antimicrobial agents in cosmetic emulsion. *Indian Journal of Microbiology* 53(2), 232–237.

Hernández-García, P.A., Orzuna-Orzuna, J.F., Godina-Rodríguez, J.E., Chay-Canul, A.J. and Silva, G.V. (2024) A meta-analysis of essential oils as a dietary additive for weaned piglets: Growth performance, antioxidant status, immune response, and intestinal morphology. *Research in Veterinary Science* 170, 105181.

Hiltunen, T., Cairns, J., Frickel, J., Jalasvuori, M., Laakso, J. *et al.* (2018) Dual-stressor selection alters eco-evolutionary dynamics in experimental communities. *Nature Ecology & Evolution* 2, 1974–1981.

Holopainen, J. (2004) Multiple functions of inducible plant volatiles. *Trends in Plant Science* 9(11), 529–533.

Holzinger, M. (2013) *Paracelsus: Das Buch Paragranum: Septem Defensiones*. Holzinger, Berlin.

Hotchkiss, S.A.M. (1998) Absorption of fragrance ingredients using in vitro models with human skin. In: Frosch, P.J., Johansen, J.D. and White, I.R. (eds) *Fragrances*. Springer, Berlin and Heidelberg, Germany, pp. 125–135.

Huang, X., Li, H., Ruan, Y., Li, Z., Yang, H. *et al.* (2022) An integrated approach utilizing raman spectroscopy and chemometrics for authentication and detection of adulteration of agarwood essential oils. *Frontiers in Chemistry* 10, 1036082.

Hudz, N., Kobylinska, L., Pokajewicz, K., Horčinová Sedláčková, V., Fedin, R. *et al.* (2023) *Mentha piperita*: Essential oil and extracts, their biological activities, and perspectives on the development of new medicinal and cosmetic products. *Molecules (Basel, Switzerland)* 28, 7444.

Ibrahim, S.S. (2020) Essential oil nanoformulations as a novel method for insect pest control in horticulture. In: Kossi Baimey, H., Hamamouch, N. and Adjiguita Kolombia, Y. (eds) *Horticultural Crops*. IntechOpen, pp. 195–208.

Inarejos-Garcia, A.M., Heil, J., Martorell, P., Álvarez, B., Llopis, S. *et al.* (2023) Effect-directed, chemical and taxonomic profiling of peppermint proprietary varieties and corresponding leaf extracts. *Antioxidants (Basel, Switzerland)* 12(2), 476.

Industrieverband Körperpflege- und Waschmittel e. V (2023) Bericht Nachhaltigkeit in der Wasch-, Pflege- und Reinigungsmittelbranche in Deutschland - Ausgabe 2023. Available at: https://www.ikw.org/fileadmin/IKW_Dateien/downloads/Haushaltspflege/2023_IKW_Nachhaltigkeitsbericht.pdf (accessed 9 April 2025).

International Cat Care (n.d.) Inhaler Training: Training Cats for Comfort with Inhaled Therapy. Available at: https://icatcare.org/articles/asthma-and-chronic-bronchitis-in-cats (accessed 12 November 2024).

International Fragrance Association (2008) Information Letter 799 - furocoumarins in finished cosmetic products.

International Fragrance Association (2023) Index of IFRA standards - 51st amendment. Available at: https://ifrafragrance.org/safe-use/standards-documentation (accessed 13 November 2024).

International Organization for Standardization (2021) ISO 9235:2021(en) Aromatic Natural Raw Materials — Vocabulary. Available at: https://www.iso.org/obp/ui/en/#iso:std:iso:9235:ed-3:v1:en (accessed 11 November 2024).

International Organization for Standardization (2023) ISO 210:2023(en) Essential Oils — General Requirements and Guidelines for Packaging, Conditioning and Storage. Available at: https://www.iso.org/obp/ui/en/#iso:std:iso:210:ed-1:v1:en (accessed 11 November 2024).

Ipek, E., Zeytinoglu, H., Okay, S., Tuylu, B.A., Kurkcuoglu, M. *et al.* (2005) Genotoxicity and antigenotoxicity of *Origanum* oil and carvacrol evaluated by Ames Salmonella/microsomal test. *Food Chemistry* 93(3), 551–556.

Itani, W.S., El-Banna, S.H., Hassan, S.B., Larsson, R.L., Bazarbachi, A. *et al.* (2008) Anti colon cancer components from Lebanese sage (*Salvia libanotica*) essential oil: Mechanistic basis. *Cancer Biology & Therapy* 7, 1765–1773.

Jain, S., Arora, P. and Nainwal, L.M. (2022) Essential oils as potential source of anti-dandruff agents: A review. *Combinatorial Chemistry & High Throughput Screening* 25(9), 1411–1426.

Jalševac, F., Terra, X., Rodríguez-Gallego, E., Beltran-Debón, R., Blay, M.T. *et al.* (2022) The hidden one: What we know about bitter taste receptor 39. *Frontiers in Endocrinology* 13, 854718.

Jamshidi-Kia, F., Lorigooini, Z. and Amini-Khoei, H. (2018) Medicinal plants: Past history and future perspective. *Journal of Herbmed Pharmacology* 7(1), 1–7.

Janssens, L., de Pooter, H.L., Schamp, N.M. and Vandamme, E.J. (1992) Production of flavours by microorganisms. *Process Biochemistry* 27(4), 195–215.

Javid, A., Raza, Z.A., Hussain, T. and Rehman, A. (2014) Chitosan microencapsulation of various essential oils to enhance the functional properties of cotton fabric. *Journal of microencapsulation* 31, 461–468.

Johansen, J.D. (2003) Fragrance contact allergy: A clinical review. *American Journal of Clinical Dermatology* 4(11), 789–798.

Jork, H., Funk, W., Fischer, W. and Wimmer, H. (1989) *Dünnschicht-Chromatographie: Reagenzien u Nachweismethoden*. VCH, Weinheim, Germany.

Jung, D.-J., Cha, J.-Y., Kim, S.-E., Ko, I.-G. and Jee, Y.-S. (2013) Effects of ylang-ylang aroma on blood pressure and heart rate in healthy men. *Journal of Exercise Rehabilitation* 9(2), 250–255.

Kada, T. and Shimoi, K. (1987) Desmutagens and bio-antimutagens-their modes of action. *BioEssays* 7(3), 113–116.

Kalbe, B., Knobloch, J., Schulz, V.M., Wecker, C., Schlimm, M. *et al.* (2016) Olfactory receptors modulate physiological processes in human airway smooth muscle cells. *Frontiers in Physiology* 7, 339.

Kamal, F.Z., Stanciu, G.D., Lefter, R., Cotea, V.V., Niculaua, M. *et al.* (2022) Chemical composition and antioxidant activity of *Ammi visnaga* L. essential oil. *Antioxidants (Basel, Switzerland)* 11(2), 347.

Karalis, V., Macheras, P., van Peer, A. and Shah, V.P. (2008) Bioavailability and bioequivalence: Focus on physiological factors and variability. *Pharmaceutical Research* 25, 1956–1962.

Karlberg, A.-T., Börje, A., Duus Johansen, J., Lidén, C., Rastogi, S. *et al.* (2013) Activation of non-sensitizing or low-sensitizing fragrance substances into potent sensitizers - prehaptens and prohaptens. *Contact Dermatitis* 69(6), 323–334.

Karpouhtsis, I., Pardali, E., Feggou, E., Kokkini, S., Scouras, Z.G. *et al.* (1998) Insecticidal and genotoxic activities of oregano essential oils. *Journal of Agricultural and Food Chemistry* 46, 1111–1115.

Katsoulos, P.D., Karatzia, M.A., Dovas, C.I., Filioussis, G., Papadopoulos, E. *et al.* (2017) Evaluation of the in-field efficacy of oregano essential oil administration on the control of neonatal diarrhea syndrome in calves. *Research in Veterinary Science* 115, 478–483.

Kejlová, K., Jírová, D., Bendová, H., Gajdoš, P. and Kolářová, H. (2010) Phototoxicity of essential oils intended for cosmetic use. *Toxicology In Vitro* 24, 2084–2089.

Kerscher, M. and Buntrock, H. (2011) Update on cosmeceuticals. *Journal of the German Society of Dermatology* 9(4), 314–326.

Khalil, N., Bishr, M., Desouky, S. and Salama, O. (2020) *Ammi visnaga* L., a potential medicinal plant: A review. *Molecules (Basel, Switzerland)* 25, 301.

Kharbach, M., Marmouzi, I., El Jemli, M., Bouklouze, A. and Vander Heyden, Y. (2020) Recent advances in untargeted and targeted approaches applied in herbal-extracts and essential-oils fingerprinting - a review. *Journal of Pharmaceutical and Biomedical Analysis* 177, 112849.

Khattak, F., Ronchi, A., Castelli, P. and Sparks, N. (2014) Effects of natural blend of essential oil on growth performance, blood biochemistry, cecal morphology, and carcass quality of broiler chickens. *Poultry Science* 93(1), 132–137.

Khodabakhsh, P., Shafaroodi, H. and Asgarpanah, J. (2015) Analgesic and anti-inflammatory activities of *Citrus aurantium* L. blossoms essential oil (neroli): Involvement of the nitric oxide/cyclic-guanosine monophosphate pathway. *Journal of Natural Medicines* 69(3), 324–331.

Kissels, W., Wu, X. and Santos, R.R. (2017) Short communication: Interaction of the isomers carvacrol and thymol with the antibiotics doxycycline and tilmicosin: *In vitro* effects against pathogenic bacteria commonly found in the respiratory tract of calves. *Journal of Dairy Science* 100, 970–974.

Klenk, F.K. and Schulz, B. (2022) Inhalative Therapie chronischer Erkrankungen der unteren Atemwege bei Hund und Katze – eine Literaturübersicht. *Tierarztliche Praxis. Ausgabe K, Kleintiere/Heimtiere* 50(4), 279–292.

Kontaris, I., East, B.S. and Wilson, D.A. (2020) Behavioral and neurobiological convergence of odor, mood and emotion: A review. *Frontiers in Behavioral Neuroscience* 14, 35.

Kucharska, M., Frydrych, B., Wesolowski, W., Szymanska, J.A. and Kilanowicz, A. (2021) A comparison of the composition of selected commercial sandalwood oils with the international standard. *Molecules (Basel, Switzerland)* 26, 2249.

Kuo, M.L., Lee, K.C. and Lin, J.K. (1992) Genotoxicities of nitropyrenes and their modulation by apigenin, tannic acid, ellagic acid and indole-3-carbinol in the Salmonella and CHO systems. *Mutation Research* 270(2), 87–95.

Lafhal, S., Bombarda, I., Dupuy, N., Jean, M., Ruiz, K. *et al.* (2020) Chiroptical fingerprints to characterize lavender and lavandin essential oils. *Journal of Chromatography: A* 1610, 460568.

Lal, R.K., Gupta, P., Chanotiya, C.S. and Sarkar, S. (2018) Traditional plant breeding in *Ocimum*. In: Shasany, A.K. and Kole, C. (eds) *The Ocimum Genome*. Springer International Publishing, Cham, Switzerland, pp. 89–98.

Lammari, N., Louaer, O., Meniai, A.H., Fessi, H. and Elaissari, A. (2021) Plant oils: From chemical composition to encapsulated form use. *International Journal of Pharmaceutics* 601, 120538.

Lang, T., Di Pizio, A., Risso, D., Drayna, D. and Behrens, M. (2023) Activation profile of TAS2R2, the 26th human bitter taste receptor. *Molecular Nutrition & Food Research* 67(11), e2200775.

Langeveld, W.T., Veldhuizen, E.J. and Burt, S.A. (2014) Synergy between essential oil components and antibiotics: A review. *Critical Reviews in Microbiology* 40(1), 76–94.

Lass-Flörl, C. and Mayr, A. (2007) Human protothecosis. *Clinical Microbiology Reviews* 20(2), 230–242.

Lazutka, J.R., Mierauskiene, J., Slapsyte, G. and Dedonyte, V. (2001) Genotoxicity of dill (*Anethum graveolens* L.), peppermint (*Mentha* × *piperita* L.) and pine (*Pinus sylvestris* L.) essential oils in human lymphocytes and *Drosophila melanogaster*. *Food and Chemical Toxicology* 39, 485–492.

Lebanov, L., Ghiasvand, A. and Paull, B. (2021) Data handling and data analysis in metabolomic studies of essential oils using GC-MS. *Journal of Chromatography: A* 1640, 461896.

LeBel, G., Vaillancourt, K., Bercier, P. and Grenier, D. (2019) Antibacterial activity against porcine respiratory bacterial pathogens and in vitro biocompatibility of essential oils. *Archives of Microbiology* 201, 833–840.

Lebensmittel- und Futtermittelgesetzbuch (2021) German Food and Feed Act in the version published on 15 September 2021 (BGBl. I p.4253). Available at: https://www.gesetze-im-internet.de/lfgb/LFGB.pdf (accessed 11 November 2024).

Leijon, S.C.M., Neves, A.F., Breza, J.M., Simon, S.A., Chaudhari, N. *et al.* (2019) Oral thermosensing by murine trigeminal neurons: Modulation by capsaicin, menthol and mustard oil. *The Journal of Physiology* 597, 2045–2061.

Lertsatitthanakorn, P., Taweechaisupapong, S., Aromdee, C. and Khunkitti, W. (2006) In vitro bioactivities of essential oils used for acne control. *International Journal of Aromatherapy* 16(1), 43–49.

Levey, M. (1959) Chemistry and chemical technology in ancient Mesopotamia. Elsevier, Amsterdam.

Levorato, S., Dominici, L., Fatigoni, C., Zadra, C., Pagiotti, R. *et al.* (2018) In vitro toxicity evaluation of estragole-containing preparations derived from *Foeniculum vulgare* Mill. (fennel) on HepG2 cells. *Food and Chemical Toxicology* 111, 616–622.

Li, P. and Liu, J. (2024) Micro(nano)plastics in the human body: Sources, occurrences, fates, and health risks. *Environmental Science & Technology* 58, 3065–3078.

Liang, Z., Zhang, P. and Fang, Z. (2020) Modern technologies for extraction of aroma compounds from fruit peels: A review. *Critical Reviews in Food Science and Nutrition* 62, 1284–1307.

Lillehei, A.S. and Halcon, L.L. (2014) A systematic review of the effect of inhaled essential oils on sleep. *Journal of Alternative and Complementary Medicine (NewYork, N.Y.)* 20(6), 441–451.

Lim, X.-Y., Li, J., Yin, H.-M., He, M., Li, L. *et al.* (2023) Stabilization of essential oil: Polysaccharide-based drug delivery system with plant-like structure based on biomimetic concept. *Polymers* 15, 3338.

Lin, L., Long, N., Qiu, M., Liu, Y., Sun, F. *et al.* (2021) The inhibitory efficiencies of geraniol as an anti-inflammatory, antioxidant, and antibacterial, natural agent against methicillin-resistant *Staphylococcus aureus* infection *in vivo*. *Infection and Drug Resistance* 14, 2991–3000.

Linh, N.T., Qui, N.H. and Triatmojo, A. (2022) The effect of nano-encapsulated herbal essential oils on poultry' s health. *Archives of Razi Institute* 77, 2013–2021.

Liu, H., Wang, J., He, T., Becker, S., Zhang, G. *et al.* (2018) Butyrate: A double-edged sword for health? *Advances in Nutrition (Bethesda, Md.)* 9(1), 21–29.

Liu, S., Sun, H., Ma, G., Zhang, T., Wang, L. *et al.* (2022) Insights into flavor and key influencing factors of Maillard reaction products: A recent update. *Frontiers in Nutrition* 9, 973677.

Lopes, T.S., Fontoura, P.S., Oliveira, A., Rizzo, F.A., Silveira, S. *et al.* (2020) Use of plant extracts and essential oils in the control of bovine mastitis. *Research in Veterinary Science* 131, 186–193.

Loreto, F. and D'Auria, S. (2022) How do plants sense volatiles sent by other plants? *Trends in Plant Science* 27(1), 29–38.

Loreto, F., Bagnoli, F. and Fineschi, S. (2009) One species, many terpenes: Matching chemical and biological diversity. *Trends in Plant Science* 14(8), 416–420.

Löscher, W., Potschka, H. and Richter, A. (eds) (2014) *Pharmakotherapie bei Haus- und Nutztieren: Begründet von W. Löscher, F.R. Ungemach und R. Kroker*, 9th edn. Enke, Stuttgart, Germany.

Louw, S. (2021) Recent trends in the chromatographic analysis of volatile flavor and fragrance compounds: Annual review 2020. *Analytical Science Advances* 2(3–4), 157–170.

Maccioni, A.M., Anchisi, C., Sanna, A., Sardu, C. and Dessì, S. (2002) Preservative systems containing essential oils in cosmetic products. *International Journal of Cosmetic Science* 24(1), 53–59.

Maggio, F., Rossi, C., Serio, A., Chaves-Lopez, C., Casaccia, M. *et al.* (2025) Anti-biofilm mechanisms of action of essential oils by targeting genes involved in quorum sensing, motility, adhesion, and virulence: A review. *International Journal of Food Microbiology* 426, 110874.

Mahanta, B.P., Sut, D., Kemprai, P., Paw, M., Lal, M. *et al.* (2020) A 1 H-NMR spectroscopic method for the analysis of thermolabile chemical markersfrom the essential oil of black turmeric (curcuma caesia) rhizome: applicationin post-harvest analysis. *Phytochemical Analysis* 31(1), 28–36.

Mahanta, B.P., Bora, P.K., Kemprai, P., Borah, G., Lal, M. *et al.* (2021) Thermolabile essential oils, aromas and flavours: Degradation pathways, effect of thermal processing and alteration of sensory quality. *Food Research International* 145, 110404.

Mahesh, S.K., Fathima, J. and Veena, V.G. (2019) Cosmetic potential of natural products: Industrial applications. In: Swamy, M.K. and Akhtar, M.S. (eds) *Natural Bio-Active Compounds*. Springer Singapore, Singapore, pp. 215–250.

Majeed, H., Bian, Y.-Y., Ali, B., Jamil, A., Majeed, U. *et al.* (2015) Essential oil encapsulations: Uses, procedures, and trends. *RSC Advances* 5, 58449–58463.

Makarska-Białokoz, M. (2020) History and significance of phytotherapy in the human history: 3. The development of phytotherapy from the Middle ages to modern times. *Archives of Physiotherapy & Global Researches* 24(2), 17–22.

Malnic, B., Godfrey, P.A. and Buck, L.B. (2004) The human olfactory receptor gene family. *Proceedings of the National Academy of Sciences of the United States of America* 101, 2584–2589.

Mancianti, F. and Ebani, V.V. (2020) Biological activity of essential oils. *Molecules (Basel, Switzerland)* 25(678).

Manion, C.R. and Widder, R.M. (2017) Essentials of essential oils. *American Journal of Health-System Pharmacy* 74(9), e153–e162.

Marques, H.M.C. (2010) A review on cyclodextrin encapsulation of essential oils and volatiles. *Flavour and Fragrance Journal* 25(5), 313–326.

Martin, M.J., Trujillo, L.A., Garcia, M.C., Alfaro, M.C. and Muñoz, J. (2018a) Effect of emulsifier HLB and stabilizer addition on the physical stability of thyme essential oil emulsions. *Journal of Dispersion Science and Technology* 39, 1627–1634.

Martin, S.F., Rustemeyer, T. and Thyssen, J.P. (2018b) Recent advances in understanding and managing contact dermatitis. *F1000Research* 7.

Martins, E., Poncelet, D., Rodrigues, R.C. and Renard, D. (2017) Oil encapsulation techniques using alginate as encapsulating agent: Applications and drawbacks. *Journal of Microencapsulation* 34, 754–771.

Masango, P. (2005) Cleaner production of essential oils by steam distillation. *Journal of Cleaner Production* 13, 833–839.

Masotti, V., Juteau, F., Bessière, J.M. and Viano, J. (2003) Seasonal and phenological variations of the essential oil from the narrow endemic species *Artemisia molinieri* and its biological activities. *Journal of Agricultural and Food Chemistry* 51, 7115–7121.

Matté, E.H.C., Luciano, F.B. and Evangelista, A.G. (2023) Essential oils and essential oil compounds in animal production as antimicrobials and anthelmintics: An updated review. *Animal Health Research Reviews* 24(1), 1–11.

Mazur, M., Ndokaj, A., Bietolini, S., Nisii, V., Duś-Ilnicka, I. *et al.* (2022) Green dentistry: Organic toothpaste formulations. A literature review. *Dental and Medical Problems* 59(3), 461–474.

McMullen, R.L. and Dell'Acqua, G. (2023) History of natural ingredients in cosmetics. *Cosmetics* 10(3), 71.

Mehta, P., Bothiraja, C., Mahadik, K., Kadam, S. and Pawar, A. (2018) Phytoconstituent based dry powder inhalers as biomedicine for the management of pulmonary diseases. *Biomedicine & Pharmacotherapy* 108, 828–837.

Menoud, V., Holinger, M., Graf-Schiller, S., Mayer, P., Gerber, L. *et al.* (2024) Comparison between intrauterine application of an antibiotic and an herbal product to treat clinical endometritis in dairy cattle - a randomized multicentre field study. *Research in Veterinary Science* 172, 105250.

Meyer, U. (2004) Verträglichkeit natürlicher ätherischer Öle bei ausgewiesenen Duftstoff-Mix-Allergikern. *Der Merkurstab* 57(1), 51–53.

Michaleas, S.N., Laios, K., Tsoucalas, G. and Androutsos, G. (2021) Theophrastus Bombastus Von Hohenheim (Paracelsus) (1493-1541): The eminent physician and pioneer of toxicology. *Toxicology Reports* 8, 411–414.

Mohammed Aggad, F.Z., Ilias, F., Elghali, F., Mrabet, R., El Haci, I.A. *et al.* (2025) Evaluation of antibacterial activity in some Algerian essential oils and selection of *Thymus vulgaris* as a potential biofilm and quorum sensing inhibitor against Pseudomonas aeruginosa. *Chemistry & Biodiversity*.

Mohammed, N.K., Tan, C.P., Manap, Y.A., Muhialdin, B.J. and Hussin, A.S.M. (2020) Spray drying for the encapsulation of oils – a review. *Molecules* 25, 3873.

Moore, A., Ankney, E., Swor, K., Poudel, A., Satyal, P. *et al.* (2025) Leaf essential oil compositions and enantiomeric distributions of monoterpenoids in *Pinus* species: *Pinus albicaulis, Pinus flexilis, Pinus lambertiana, Pinus monticola,* and *Pinus sabiniana. Molecules (Basel, Switzerland)* 30, 244.

Móricz, Á.M., Häbe, T.T., Böszörményi, A., Ott, P.G. and Morlock, G.E. (2015) Tracking and identification of antibacterial components in the essential oil of *Tanacetum vulgare* L. by the combination of high-performance thin-layer chromatography with direct bioautography and mass spectrometry. *Journal of Chromatography: A* 1422, 310–317.

Morlock, G.E. (2021) High-performance thin-layer chromatography combined with effect-directed assays and high-resolution mass spectrometry as an emerging hyphenated technology: A tutorial review. *Analytica Chimica Acta* 1180, 338644.

Morlock, G.E. (2022) Planar chromatographic super-hyphenations for rapid dereplication. *Phytochemistry Reviews* 24, 1–12.

Morlock, G.E. and Meyer, D. (2023) Designed genotoxicity profiling detects genotoxic compounds in staple food such as healthy oils. *Food Chemistry* 408, 135253.

Morlock, G.E. and Heil, J. (2025, in press) Fast un masking hazards of *safe* perfumes. *Journal of Chromatography: A.*

Morlock, G.E., Koch, J. and Schwack, W. (2023) Miniaturized open-source 2LabsToGo screening of lactose-free dairy products and saccharide-containing foods. *Journal of Chromatography: A* 1688, 463720.

Muhoza, B., Xia, S., Wang, X., Zhang, X., Li, Y. *et al.* (2022) Microencapsulation of essential oils by complex coacervation method: Preparation, thermal stability, release properties and applications. *Critical Reviews in Food Science and Nutrition* 62, 1363–1382.

Mukurumbira, A.R., Shellie, R.A., Keast, R., Palombo, E.A. and Jadhav, S.R. (2022) Encapsulation of essential oils and their application in antimicrobial active packaging. *Food Control* 136, 108883.

Müller, E. and Schneider, M. (2015) Ätherische Öle gezielt auswählen: Was leistet das Aromatogramm? Wann wird es eingesetzt? *Zeitschrift für Phytotherapie* 36(Suppl.1).

Müller, I., Gulde, A. and Morlock, G.E. (2023) Bioactive profiles of edible vegetable oils determined using 10D hyphenated comprehensive high-performance thin-layer chromatography (HPTLC×HPTLC) with on-surface metabolism (nanoGIT) and planar bioassays. *Frontiers in Nutrition* 10, 1227546.

Mulyaningsih, S., Sporer, F., Zimmermann, S., Reichling, J. and Wink, M. (2010) Synergistic properties of the terpenoids aromadendrene and 1,8-cineole from the essential oil of *Eucalyptus globulus* against antibiotic-susceptible and antibiotic-resistant pathogens. *Phytomedicine: International Journal of Phytotherapy and Phytopharmacology* 17, 1061–1066.

Murphy, B.J., Wilson, T.M., Ziebarth, E.A., Bowerbank, C.R. and Carlson, R.E. (2024) Authentication of fennel, star anise, and anise essential oils by gas chromatography (GC/MS) and stable isotope ratio (GC/IRMS) analyses 13(2), 214.

Muyima, N.Y.O., Zulu, G., Bhengu, T. and Popplewell, D. (2002) The potential application of some novel essential oils as natural cosmetic preservatives in an aqueous cream formulation. *Flavour and Fragrance Journal* 17(4), 258–266.

Nahrstedt, A. and Butterweck, V. (2010) Lessons learned from herbal medicinal products: The example of St.John's Wort (perpendicular). *Journal of Natural Products* 73, 1015–1021.

Nakajima, D., Yamachi, M., Misaka, S., Shimomura, K. and Maejima, Y. (2024) Sex differences in the effects of aromatherapy on anxiety and salivary oxytocin levels. *Frontiers in Endocrinology* 15, 1380779.

Naquvi, K.J., Ansari, S.H., Ali, M. and Najmi, A. (2014) Volatile oil composition of *Rosa damascena* Mill. (Rosaceae). *Journal of Pharmacognosy and Phytochemistry* 5, 130–134.

Nascimento, J.C., Gonçalves, V.S.D.S., Souza, B.R.S., Nascimento, L. d. C., de Carvalho, B.M.R. *et al.* (2024) Effectiveness of aromatherapy with sweet orange oil (*Citrus sinensis* L.) in relieving pain and anxiety during labor. *Explore (New York, N.Y.)* 21(1), 103081.

NATRUE (2021) NATRUE's consumer study on consumer perception about brands and seals in regard to cosmetics in Germany and France. Available at: https://www.natrue.org/uploads/2021/03/NATRUE_Consumer-study_DE-and-FR_2021.pdf (accessed 13 2024).

Natsch, A., Nägelin, M., Leijs, H., van Strien, M., Giménez-Arnau, E. *et al.* (2019) Exposure source for skin sensitizing hydroperoxides of limonene and linalool remains elusive: An analytical market surveillance. *Food and Chemical Toxicology* 127, 156–162.

Nayebi, N., Khalili, N., Kamalinejad, M. and Emtiazy, M. (2017) A systematic review of the efficacy and safety of *Rosa damascena* Mill. with an overview on its phytopharmacological properties. *Complementary Therapies in Medicine* 34, 129–140.

Nazzaro, F., Fratianni, F., de Martino, L., Coppola, R. and de Feo, V. (2013) Effect of essential oils on pathogenic bacteria. *Pharmaceuticals (Basel, Switzerland)* 6, 1451–1474.

Nerín, C., Bourdoux, S., Faust, B., Gude, T., Lesueur, C. *et al.* (2022) Guidance in selecting analytical techniques for identification and quantification of nonintentionally added substances (NIAS) in food contact materials (FCMS). *Food Additives & Contaminants: Part A* 39(3), 620–643.

Nestmann, E.R. and Lee, E.G. (1983) Mutagenicity of constituents of pulp and paper mill effluent in growing cells of *Saccharomyces cerevisiae*. *Mutation Research* 119(3), 273–280.

Newman, J.D. and Chappell, J. (1999) Isoprenoid biosynthesis in plants: Carbon partitioning within the cytoplasmic pathway. *Critical Reviews in Biochemistry and Molecular Biology* 34(2), 95–106.

Nguyen, D.A., Muhammad, M.K. and Lee, G.L. (2020) Phytophotodermatitis. In: Trevino, J. and Chen, A.Y.-Y. (eds) *Dermatological Manual of Outdoor Hazards*. Springer International Publishing, Cham, Switzerland, pp. 43–56.

Ni, M. (1995) *The Yellow Emperor's Classic of Medicine: A New Translation of the Neijing Suwen with Commentary*. Shambhala, Boston.

Niedermaier, G. and Gehlen, H. (2009) Möglichkeiten der Inhalationstherapie zur Behandlung der chronisch obstruktiven Bronchitis des Pferdes. *Pferdeheilkunde* 25, 327–332.

Novgorodov, S.A. and Gudz, T.I. (1996) Permeability transition pore of the inner mitochondrial membrane can operate in two open states with different selectivities. *Journal of Bioenergetics and Biomembranes* 28(2), 139–146.

Nurzyńska-Wierdak, R., Pietrasik, D. and Walasek-Janusz, M. (2022) Essential oils in the treatment of various types of acne - a review. *Plants (Basel, Switzerland)* 12(1), 90.

Ogueta, I.A., Brared Christensson, J., Giménez-Arnau, E., Brans, R., Wilkinson, M. *et al.* (2022) Limonene and linalool hydroperoxides review: Pros and cons for routine patch testing. *Contact Dermatitis* 87(1), 1–12.

Ohashi, T., Miyazawa, Y., Ishizaki, S., Kurobayashi, Y. and Saito, T. (2019) Identification of odor-active trace compounds in blooming flower of damask rose (*Rosa damascena*). *Journal of Agricultural and Food Chemistry* 67, 7410–7415.

Oladipupo, S.O., Hu, X.P. and Appel, A.G. (2022) Essential oils in urban insect management - a review. *Journal of Economic Entomology* 115, 1375–1408.

Oliveira, E.C.V. de, Salvador, D.S., Holsback, V., Shultz, J.D., Michniak-Kohn, B.B. *et al.* (2021) Deodorants and antiperspirants: identification of new strategies and perspectives to prevent and control malodor and sweat of the body. *International Journal of Dermatology* 60(5), 613–619.

Palhares Campolina, J., Gesteira Coelho, S., Belli, A.L., Samarini Machado, F., Pereira, L.G.R. *et al.* (2021) Effects of a blend of essential oils in milk replacer on performance, rumen fermentation, blood parameters, and health scores of dairy heifers. *PloS one* 16(3), e0231068.

Pandey, V.K., Islam, R.U., Shams, R. and Dar, A.H. (2022) A comprehensive review on the application of essential oils as bioactive compounds in nano-emulsion based edible coatings of fruits and vegetables. *Applied Food Research* 2(1), 100042.

Parente, E. and Ares, G. (2021) How do appearance and fragrance influence expectations and conceptual associations of cosmetic products? An exploratory case study with liquid hand soap. *Journal of Sensory Studies* 36(2), e12637.

Pavoni, L., Perinelli, D.R., Bonacucina, G., Cespi, M. and Palmieri, G.F. (2020) An overview of micro- and nanoemulsions as vehicles for essential oils: Formulation, preparation and stability. *Nanomaterials (Basel, Switzerland)* 10(1), 135.

Pedonese, F., Longo, E., Torracca, B., Najar, B., Fratini, F. *et al.* (2022) Antimicrobial and anti-biofilm activity of manuka essential oil against *Listeria monocytogenes* and *Staphylococcus aureus* of food origin. *Italian Journal of Food Safety* 11(1), 10039.

Peters, R.J., Groeneveld, I., Sanchez, P.L., Gebbink, W., Gersen, A. *et al.* (2019) Review of analytical approaches for the identification of non-intentionally added substances in paper and board food contact materials. *Trends in Food Science & Technology* 85, 44–54.

Phanse, S.K. and Chandra, S. (2024) Spray drying encapsulation of essential oils: Insights on various factors affecting the physicochemical properties of the microcapsules. *Flavour and Fragrance Journal* 39(2), 93–115.

Phi, N.T.L., van Hung, P., Chi, P.T.L. and Dung, N.H. (2015) Impact of extraction methods on antioxidant and antimicrobial activities of citrus essential oils. *Journal of Essential Oil Bearing Plants* 18(4), 806–817.

Phucharoenrak, P. and Trachootham, D. (2024) Bergaptol, a major furocoumarin in citrus: Pharmacological properties and toxicity. *Molecules (Basel, Switzerland)* 29, 713.

Pibiri, M., Goel, A., Vahekeni, N. and Roulet, C. (2006) Indoor air purification and ventilation systems sanitation with essential oils. *International Journal of Aromatherapy* 16(3–4), 149–153.

Pires, J.B., Santos, F.N.D., Costa, I.H.d.L, Kringel, D.H., Da Zavareze, E.R. *et al.* (2023) Essential oil encapsulation by electrospinning and electrospraying using food proteins: A review. *Food Research International* 170, 112970.

Plant, R.M., Dinh, L., Argo, S. and Shah, M. (2019) The essentials of essential oils. *Advances in Pediatrics* 66, 111–122.

Pokajewicz, K., Białoń, M., Svydenko, L., Hudz, N., Balwierz, R. *et al.* (2022) Comparative evaluation of the essential oil of the new Ukrainian *Lavandula angustifolia* and *Lavandula* x *intermedia* cultivars grown on the same plots. *Molecules (Basel, Switzerland)* 27, 2152.

Poudel, D.K., Rokaya, A., Ojha, P.K., Timsina, S., Satyal, R. *et al.* (2021) The chemical profiling of essential oils from different tissues of *Cinnamomum camphora* L. and their antimicrobial activities. *Molecules* 26(17), 5132.

Pruthi, J.S. (1998) *Quality Assurance in Spices and Spice Products. Modern Methods of Analysis.* Allied Publishers Limited, New Delhi.

Pybus, D.H. (2006) The history of aroma chemistry and perfume. In: Sell, C.S. (ed.) *Chemistry of Fragrances.* The Royal Society of Chemistry, Cambridge, UK, pp. 3–23.

Raeber, J., Favrod, S. and Steuer, C. (2023) Determination of major, minor and chiral components as quality and authenticity markers of *Rosa damascena* oil by GC–FID. *Plants (Basel, Switzerland)* 12(3), 506.

Rahmi, D., Yunilawati, R., Jati, B.N., Setiawati, I., Riyanto, A. *et al.* (2021) Antiaging and skin irritation potential of four main Indonesian essential oils. *Cosmetics* 8(4), 94.

Ramadan, M., Goeters, S., Watzer, B., Krause, E., Lohmann, K. *et al.* (2006) Chamazulene carboxylic acid and matricin: A natural profen and its natural prodrug, identified through similarity to synthetic drug substances. *Journal of Natural Products* 69, 1041–1045.

Ramel, C., Alekperov, U.K., Ames, B.N., Kada, T. and Wattenberg, L.W. (1986) International commission for protection against environmental mutagens and carcinogens. ICPEMC publication no.12. inhibitors of mutagenesis and their relevance to carcinogenesis. Report by ICPEMC Expert Group on Antimutagens and Desmutagens. *Mutation Research* 168(1), 47–65.

Ramezani, S., Saharkhiz, M.J., Ramezani, F. and Fotokian, M.H. (2008) Use of essential oils as bioherbicides. *Journal of Essential Oil Bearing Plants* 11(3), 319–327.

Ranasinghe, S., Armson, A., Lymbery, A.J., Zahedi, A. and Ash, A. (2023) Medicinal plants as a source of antiparasitics: An overview of experimental studies. *Pathogens and Global Health* 117(6), 535–553.

Rasheed, D.M., Serag, A., Abdel Shakour, Z.T. and Farag, M. (2021) Novel trends and applications of multidimensional chromatography in the analysis of food, cosmetics and medicine bearing essential oils. *Talanta* 223(Pt 1), 121710.

Regnault-Roger, C., Vincent, C. and Arnason, J.T. (2012) Essential oils in insect control: Low-risk products in a high-stakes world. *Annual Review of Entomology* 57(1), 405–424.

Reich, E. and Schibli, A. (eds) (2011) *High-Performance Thin-Layer Chromatography for the Analysis of Medicinal Plants*. Thieme, New York.

Reichling, J. (2016) *Heilpflanzenkunde für die Veterinärpraxis*, 3rd edn. Springer Berlin Heidelberg, Berlin and Heidelberg, Germany.

Reineccius, G.A. (2004) The spray drying of food flavors. *Drying Technology* 22(6), 1289–1324.

Reinhold, P. and Fehrenbach, H. (2003) Aerosole in Medizin und Veterinärmedizin. *Pneumologie* 57(5), 166.

Richter, C. and Schlegel, J. (1993) Mitochondrial calcium release induced by prooxidants. *Toxicology Letters* 67(1–3), 119–127.

Röll, M.F. (1866) *Lehrbuch der Arzneimittellehre für Thierärzte*, 2nd edn. Braunmüller, Vienna.

Romanenko, E.P., Domrachev, D.V. and Tkachev, A.V. (2022) Variations in essential oils from south Siberian conifers of the Pinaceae family: New data towards identification and quality control. *Chemistry & Biodiversity* 19(2), e202100755.

Romero, M., Jakob, K., Schmidt, J., Nimmerfroh, T., Schwack, W. *et al.* (2025, in print) Consolidating two laboratories into the most sustainable lab of the future: 2LabsToGo-Eco. *Analytica Chimica Acta*.

Ronzheimer, A., Schreiner, T. and Morlock, G.E. (2022) Multiplex planar bioassay detecting phytoestrogens and verified antiestrogens as sharp zones on normal phase. *Phytomedicine* 103, 154230.

Rossi, C.A.S., Grossi, S., Dell'Anno, M., Compiani, R. and Rossi, L. (2022) Effect of a blend of essential oils, bioflavonoids and tannins on *in vitro* methane production and *in vivo* production efficiency in dairy cows. *Animals* 12(6), 728.

Rothe, M. (1988) Time table about the history of production, processing and consumption of aroma-rich foods and flavourings. In: Rothe, M. (ed.) *Introduction to Aroma Research*. Springer Netherlands, Dordrecht, pp. 101–110.

Rovesti, P. (1977) Die destillation ist 5000 Jahre alt. In: *Dragoco Report*, Vol. 3. pp. 49–62.

Rowe, D.J. (ed.) (2006) *Chemistry and Technology of Flavors and Fragrances*. Blackwell and CRC Press, Oxford, UK and Boca Raton, FL. Available at: https://permalink.obvsg.at/AC04388644 (accessed 4 April 2025).

Roxo, M. and Wink, M. (2022) The use of the nematode *Caenorhabditis elegans* to study antioxidant and longevity-promoting plant secondary metabolites. In: Tiezzi, A., Ovidi, E. and Karpinski, T. (eds) *New Findings from Natural Substances*. Bentham Science Publishers, Sharjah Airport International Free Trade Zone, UAE, pp. 133–163.

Rubiolo, P., Sgorbini, B., Liberto, E., Cordero, C. and Bicchi, C. (2010) Essential oils and volatiles: Sample preparation and analysis. A review. *Flavour and Fragrance Journal* 25(5), 282–290.

Rychen, G., Aquilina, G., Azimonti, G., Bampidis, V., Bastos, M.d.L. *et al.* (2017) Safety and efficacy of an essential oil from *Origanum vulgare* subsp. *hirtum* (Link) letsw. var. Vulkan when used as a sensory additive in feed for all animal species. *EFSA Journal* 15(12), e05095.

Sabroe, R.A., Holden, C.R. and Gawkrodger, D.J. (2016) Contact allergy to essential oils cannot always be predicted from allergy to fragrance markers in the baseline series. *Contact Dermatitis* 74(4), 236–241.

Sadgrove, N.J., Padilla-González, G.F. and Phumthum, M. (2022) Fundamental chemistry of essential oils and volatile organic compounds, methods of analysis and authentication. *Plants (Basel, Switzerland)* 11(6), 789.

Sahoo, N., Manchikanti, P. and Dey, S. (2010) Herbal drugs: Standards and regulation. *Fitoterapia* 81, 462–471.

Sahoo, C., Champati, B.B., Dash, B., Jena, S., Ray, A. *et al.* (2022) Volatile profiling of *Magnolia champaca* accessions by gas chromatography mass spectrometry coupled with chemometrics. *Molecules (Basel, Switzerland)* 27, 7302.

Saifullah, M., Shishir, M.R.I., Ferdowsi, R., Tanver Rahman, M.R. and van Vuong, Q. (2019) Micro and nano encapsulation, retention and controlled release of flavor and aroma compounds: A critical review. *Trends in Food Science & Technology* 86, 230–251.

Sakkas, H. and Papadopoulou, C. (2017) Antimicrobial activity of basil, oregano, and thyme essential oils. *Journal of Microbiology and Biotechnology* 27, 429–438.

Salsabila, S.A. (2023) The uncertainty of essential oil as a cosmetic product market: A market sociology perspective. *International Journal of Social Science and Business* 7(2), 261–267.

Salústio, P.J., Miguel, M.G. and Cabral-Marques, H. (2015) Molecular (cyclodextrin) encapsulation of volatiles and essential oils. In: Mishra, M. (ed.) *Handbook of Encapsulation and Controlled Release*. CRC Press, Boca Raton, Florida, pp. 867–906.

Salzer, U.-J. and Jones, K. (1998) Legislation/toxicology. In: Ziegler, E. and Ziegler, H. (eds) *Flavourings*. Wiley, Oxford, UK, pp. 643–698.

Sanders, M. (2007) Inhalation therapy: An historical review. *Primary Care Respiratory Journal* 16(2), 71–81.

Sangwan, N.S., Farooqi, A., Shabih, F. and Sangwan, R.S. (2001) Regulation of essential oil production in plants. *Plant Growth Regulation* 34(1), 3–21.

Sarkic, A. and Stappen, I. (2018) Essential oils and their single compounds in cosmetics—a critical review. *Cosmetics* 5(1), 11.

Saroglou, V., Dorizas, N., Kypriotakis, Z. and Skaltsa, H.D. (2006) Analysis of the essential oil composition of eight *Anthemis* species from Greece. *Journal of Chromatography: A* 1104(1–2), 313–322.

Sarrou, E., Tsivelika, N., Chatzopoulou, P., Tsakalidis, G., Menexes, G. *et al.* (2017) Conventional breeding of Greek oregano (*Origanum vulgare* ssp. *hirtum*) and development of improved cultivars for yield potential and essential oil quality. *Euphytica* 213(5).

Scarpellini, E., Broeders, B., Schol, J., Santori, P., Addarii, M. *et al.* (2023) The use of peppermint oil in gastroenterology. *Current Pharmaceutical Design* 29(8), 576–583.

Schempp, C.M. (2011) Hyperforin - ein Multitalent für die Haut Antibakterielle und antiinflammatorische Eigenschaften von Hyperforin. In: *15th Annual Meeting of the Gesellschaft für Dermopharmazie e.V*, April 6, 2011.

Schempp, C.M., Müller, K.A., Winghofer, B., Schöpf, E. and Simon, J.C. (2002) Johanniskraut (Hypericum perforatum L.). Eine Pflanze mit Relevanz für die Dermatologie. *Der Hautarzt; Zeitschrift Fur Dermatologie, Venerologie, Und Verwandte Gebiete* 53(5), 316–321.

Schlittenlacher, T. (2022) Collection of empirical knowledge on the treatment of livestock with medicinal plants and natural substances in Bavaria. MSc. thesis, Ludwig Maximilian University of Munich.

Schmidt, E. (2016) Production of essential oils. In: Başer, K.H.C. and Buchbauer, G. (eds) *Handbook of Essential Oils: Science, Technology, and Applications.* CRC Press, Boca Raton, Florida, London, and New York, pp. 127–163.

Schmidt, S., Heymann, K., Melzig, M.F., Bereswill, S. and Heimesaat, M.M. (2016) Glycyrrhizic acid decreases gentamicin-resistance in vancomycin-resistant Enterococci. *Planta Medica* 82(18), 1540–1545.

Schnuch, A. and Griem, P. (2018) Fragrances as allergens. *Allergo Journal International* 27(6), 173–183.

Schreiner, T., Sauter, D., Friz, M., Heil, J. and Morlock, G.E. (2021) Is our natural food our homeostasis? Array of a thousand effect-directed profiles of 68 herbs and spices. *Frontiers in Pharmacology* 12, 755941.

Schubert, S., Geier, J., Brans, R., Heratizadeh, A., Kränke, B. *et al.* (2023) Patch testing hydroperoxides of limonene and linalool in consecutive patients - results of the IVDK 2018-2020. *Contact Dermatitis* 89(2), 85–94.

Schulte-Hubbert, R., Küpper, J.-H., Thomas, A.D. and Schrenk, D. (2020) Estragole: DNA adduct formation in primary rat hepatocytes and genotoxic potential in HepG2-CYP1A2 cells. *Toxicology* 444, 152566.

Schulz, H., Eder, G. and Heyder, J. (2003) Lung volume is a determinant of aerosol bolus dispersion. *Journal of Aerosol Medicine: The Official Journal of the International Society for Aerosols in Medicine* 16(3), 255–262.

Schwabl, H., Vennos, C. and Saller, R. (2013) Tibetische Rezepturen als pleiotrope Signaturen - Einsatz von Netzwerk-Arzneien bei Multimorbidität. *Forschende Komplementarmedizin (2006)* 20(2), 35–40.

Scientific Committee on Consumer Safety (2021) *Opinion on Methyl Salicylate (Methyl 2-Hydroxybenzoate).* SCCS/1633/21. Available at: https://health. ec.europa.eu/system/files/2022-08/sccs_o_255.pdf (accessed 13 November 2024).

Scientific Committee on Consumer Safety (2012) *Opinion on Fragrance Allergens in Cosmetic Products.* SCCS/1459/11. Available at: https://ec.europa.eu/health/ scientific_committees/consumer_safety/docs/sccs_o_102.pdf (accessed 14 November 2024).

Scott, B.R., Pathak, M.A. and Mohn, G.R. (1976) Molecular and genetic basis of furocoumarin reactions. *Mutation Research/Reviews in Genetic Toxicology* 39(1), 29–74.

Septiyanti, M., Meliana, Y. and Agustian, E. (2017) Effect of citronella essential oil fractions as oil phase on emulsion stability. *Proceedings of the 3rd International Symposium on Applied Chemistry 2017.* Jakarta, Indonesia, 23–24 October 2017, 1904(1), 20070.

Sertürner, F.A. (1806) Darstellung der reinen Mohnsäure (Opiumsäure) nebst einer Untersuchung des Opiums mit vorzüglicher Hinsicht auf einen darin neu entdeckten Stoff und die dahin gehörigen Bemerkungen. Vom Herrn Sertürner in Paderborn. *Journal der Pharmacie* 14, 47–93.

Seyyedi, S.-A., Sanatkhani, M., Pakfetrat, A. and Olyaee, P. (2014) The therapeutic effects of chamomilla tincture mouthwash on oral aphthae: A randomized clinical trial. *Journal of Clinical and Experimental Dentistry* 6(5), e535–e538.

Shahrajabian, M.H., Sun, W. and Cheng, Q. (2021) Pharmaceutical benefits and multidimensional uses of ajwain (*Trachyspermum ammi* L.). *Pharmacognosy Communications* 11(2), 138–141.

Shankel, D.M., Kuo, S., Haines, C. and Mitscher, L.A. (1993) Extracellular interception of mutagens. *Basic Life Sciences* 61, 65–74.

Sharma, N., Trikha, P., Athar, M. and Raisuddin, S. (2001) Inhibition of benzo[a] pyrene- and cyclophoshamide-induced mutagenicity by *Cinnamomum cassia*. *Mutation Research* 480–481, 179–188.

Sharmeen, J.B., Mahomoodally, F.M., Zengin, G. and Maggi, F. (2021) Essential oils as natural sources of fragrance compounds for cosmetics and cosmeceuticals. *Molecules (Basel, Switzerland)* 26, 666.

Sherry, M., Charcosset, C., Fessi, H. and Greige-Gerges, H. (2013) Essential oils encapsulated in liposomes: A review. *Journal of Liposome Research* 23(4), 268–275.

Shlosman, K., Rein, D.M., Shemesh, R., Koifman, N., Caspi, A. *et al.* (2022) Encapsulation of thymol and eugenol essential oils using unmodified cellulose: Preparation and characterization. *Polymers* 15, 95.

Shuba, Y.M. (2020) Beyond neuronal heat sensing: Diversity of TRPV1 heat-capsaicin receptor-channel functions. *Frontiers in Cellular Neuroscience* 14, 612480.

Sing, L., Schwack, W., Göttsche, R. and Morlock, G.E. (2022) 2LabsToGo—recipe for building your own chromatography equipment including biological assay and effect detection. *Analytical Chemistry* 94, 14554–14564.

Singhal, R.S., Kulkarni, P.R. and Rege, D.V. (1997) *Handbook of Indices of Food Quality and Authenticity*. Woodhead, Cambridge, UK.

Smail, H.O. (2019) The roles of genes in the bitter taste. *AIMS Genetics* 6(4), 88–97.

Smolarek, P. d. C., Esmerino, L.A., Chibinski, A.C., Bortoluzzi, M.C., Dos Santos, E.B. *et al.* (2015) *In vitro* antimicrobial evaluation of toothpastes with natural compounds. *European Journal of Dentistry* 9(4), 580–586.

Snapkow, I., Andreassen, M., Nygaard, U.C., Pieters, R., Dirven, H. *et al.* (2024) Mapping human immune responses to micro and nanoplastics: The norwegian Football Field Study. In: *International Conference on Microplastics, Nanoplastics & Human Health*, Dublin, Ireland, August 28–30, 2024, pp. 97–98.

Souza, A.G., Ferreira, R.R., Paula, L.C., Setz, L.F. and Rosa, D.S. (2020) The effect of essential oil chemical structures on Pickering emulsion stabilized with cellulose nanofibrils. *Journal of Molecular Liquids* 320, 114458.

Sousa, V.I., Parente, J.F., Marques, J.F., Forte, M.A. and Tavares, C.J. (2022) Microencapsulation of essential oils: A review. *Polymers* 14, 1730.

Stahl-Biskup, E. and Reher, G. (1987) Geschmack und Geruch. Die chemischen Sinne des Menschen und die Chemie der Geschmack-und Riechstoffe. Teil I: Geschmack. *Deutsche Apotheker Zeitung* 127, 2529.

Stammati, A., Bonsi, P., Zucco, F., Moezelaar, R., Alakomi, H.L. *et al.* (1999) Toxicity of selected plant volatiles in microbial and mammalian short-term assays. *Food and Chemical Toxicology* 37, 813–823.

Stan, D., Enciu, A.-M., Mateescu, A.L., Ion, A.C., Brezeanu, A.C. *et al.* (2021) Natural compounds with antimicrobial and antiviral effect and nanocarriers used for their transportation. *Frontiers in Pharmacology* 12, 723233.

Stather, G. and Döderlein, G. (1968) *Tierarznei-Rezepte*, 3rd edn. WVG, Stuttgart, Germany.

Staub, P.O., Casu, L. and Leonti, M. (2016) Back to the roots: A quantitative survey of herbal drugs in Dioscorides' De Materia Medica (ex Matthioli, 1568). *Phytomedicine: International Journal of Phytotherapy and Phytopharmacology* 23, 1043–1052.

Stefania, G., Vâtcă, A. and Vâtcă, S. (2017) The history and use of perfume in human civilization. *Agricultura* 130(3–4). Available at: https://journals.usamvcluj.ro/index.php/agricultura/article/view/12864 (accessed 13 November 2024).

Stiefel, C. and Stintzing, F. (2023) Endocrine-active and endocrine-disrupting compounds in food – occurrence, formation and relevance. *NFS Journal* 31, 57–92.

Straits Research (2021) *Vanilla Extract Market Size: Share & Trends Analysis Report By Nature (Organic, Synthetic), By Application (Food & Beverages, Pharmaceuticals, Cosmetics, Others), By Form (Liquid, Powder), By Distribution Channel (B2B, B2C) and By Region (North America, Europe, APAC, Middle East and Africa, LATAM)* (SRFB1270DR). Available at: https://straitsresearch.com/report/vanilla-extract-market (accessed 11 November 2024).

Su, X., Li, B., Chen, S., Wang, X., Song, H. *et al.* (2024) Pore engineering of micro/mesoporous nanomaterials for encapsulation, controlled release and variegated applications of essential oils. *Journal of Controlled Release: Official Journal of the Controlled Release Society* 367, 107–134.

Suphasomboon, T. and Vassanadumrongdee, S. (2022) Toward sustainable consumption of green cosmetics and personal care products: The role of perceived value and ethical concern. *Sustainable Production and Consumption* 33, 230–243.

Taape, T. (2014) Distilling reliable remedies: Hieronymus Brunschwig's Liber de arte distillandi (1500) between alchemical learning and craft practice. *Ambix* 61(3), 236–256.

Tanasă, F., Nechifor, M. and Teacă, C.-A. (2024) Essential oils as alternative green broad-spectrum biocides. *Plants (Basel, Switzerland)* 23, 3442.

Teh, S.-S. and Morlock, G.E. (2015) Effect-directed analysis of cold-pressed hemp, flax and canola seed oils by planar chromatography linked with (bio)assays and mass spectrometry. *Food Chemistry* 187, 460–468.

Thakur, D., Kaur, G., Puri, A. and Nanda, R. (2021) Therapeutic potential of essential oil-based microemulsions: Reviewing state-of-the-art. *Current Drug Delivery* 18(9), 1218–1233.

Timilsena, Y.P., Akanbi, T.O., Khalid, N., Adhikari, B. and Barrow, C.J. (2019) Complex coacervation: Principles, mechanisms and applications in microencapsulation. *International Journal of Biological Macromolecules* 121, 1276–1286.

Tistaert, C., Dejaegher, B. and Vander Heyden, Y. (2011) Chromatographic separation techniques and data handling methods for herbal fingerprints: A review. *Analytica Chimica Acta* 690(2), 148–161.

Tominaga, M. and Tominaga, T. (2005) Structure and function of TRPV1. *Pflugers Archiv: European Journal of Physiology* 451, 143–150.

Toxopeus, H. and Bouwmeester, H.J. (1992) Improvement of caraway essential oil and carvone production in The Netherlands. *Industrial Crops and Products* 1(2–4), 295–301.

Truzzi, E., Marchetti, L., Bertelli, D. and Benvenuti, S. (2021) Attenuated total reflectance-fourier transform infrared (ATR-FTIR) spectroscopy coupled with chemometric analysis for detection and quantification of adulteration in lavender and citronella essential oils. *Phytochemical Analysis* 32(6), 907–920.

Turek, C. and Stintzing, F.C. (2011) Application of high-performance liquid chromatography diode array detection and mass spectrometry to the analysis of characteristic compounds in various essential oils. *Analytical and Bioanalytical Chemistry* 400, 3109–3123.

Turek, C. and Stintzing, F.C. (2012) Impact of different storage conditions on the quality of selected essential oils. *Food Research International* 46, 341–353.

Turek, C. and Stintzing, F.C. (2013) Stability of essential oils: A review. *Comprehensive Reviews in Food Science and Food Safety* 12(1), 40–53.

Turek, C. and Stintzing, F.C. (2013) Stability of essential oils: A review. *Comprehensive Reviews in Food Science and Food Safety* 12(1), 40–53.

Ulland, S., Ian, E., Stranden, M., Borg-Karlson, A.-K. and Mustaparta, H. (2008) Plant volatiles activating specific olfactory receptor neurons of the cabbage moth *Mamestra brassicae* L. (Lepidoptera, Noctuidae). *Chemical Senses* 33(6), 509–522.

United States Pharmacopeia - National Formulary (USP-NF) (2023). Available at: https://www.uspnf.com/ (accessed 14 April 2025).

Valussi, M., Donelli, D., Firenzuoli, F. and Antonelli, M. (2021) Bergamot oil: Botany, production, pharmacology. *Encyclopedia* 1(1), 152–176.

van Wyk, B.-E. and Wink, M. (2015) *Phytomedicines, Herbal Drugs, and Poisons.* The University of Chicago Press, Chicago, Illinois and Royal Botanic Gardens, Kew, UK.

van Wyk, B.-E. and Wink, M. (2017) *Medicinal Plants of the World: An Illustrated Scientific Guide to Important Medicina Plants and Their Uses*, 2nd edn. CAB International, Wallingford, UK.

Varvaresou, A., Papageorgiou, S., Tsirivas, E., Protopapa, E., Kintziou, H. *et al.* (2009) Self-preserving cosmetics. *International Journal of Cosmetic Science* 31(3), 163–175.

Vercesi, A.E., Kowaltowski, A.J., Grijalba, M.T., Meinicke, A.R. and Castilho, R.F. (1997) The role of reactive oxygen species in mitochondrial permeability transition. *Bioscience Reports* 17(1), 43–52.

Von den Driesch, A. and Peters, J. (2003) *Geschichte der Tiermedizin: 5000 Jahre Tierheilkunde*. Schattauer, Stuttgart.

Wainer, J., Thomas, A., Chimhau, T. and Harding, K.G. (2022) Extraction of essential oils from *Lavandula × intermedia* 'Margaret Roberts' using steam distillation, hydrodistillation, and cellulase-assisted hydrodistillation: Experimentation and cost analysis. *Plants (Basel, Switzerland)* 11(24), 3479.

Wang, Z.-J. and Heinbockel, T. (2018) Essential oils and their constituents targeting the GABAergic system and sodium channels as treatment of neurological diseases. *Molecules (Basel, Switzerland)* 23, 1061.

Wang, M., Avula, B., Wang, Y.-H., Zhao, J., Avonto, C. *et al.* (2014) An integrated approach utilising chemometrics and GC/MS for classification of chamomile flowers, essential oils and commercial products. *Food Chemistry* 152, 391–398.

Wang, M., Zhao, J., Ali, Z., Avonto, C. and Khan, I.A. (2021) A novel approach for lavender essential oil authentication and quality assessment. *Journal of Pharmaceutical and Biomedical Analysis* 199, 114050.

Wang, M., Lee, J., Zhao, J., Chatterjee, S., Chittiboyina, A.G. *et al.* (2024) Comprehensive quality assessment of peppermint oils and commercial products:

An integrated approach involving conventional and chiral GC/MS coupled with chemometrics. *Journal of Chromatography: B* 1232, 123953.

Waters, M.D., Stack, H.F., Jackson, M.A., Brockman, H.E. and De, F.S. (1996) Activity profiles of antimutagens: In vitro and in vivo data. *Mutation Research* 350(1), 109–129.

Weisany, W., Yousefi, S., Tahir, N.A.-R., Golestanehzadeh, N., McClements, D.J. *et al.* (2022) Targeted delivery and controlled released of essential oils using nanoencapsulation: A review. *Advances in Colloid and Interface Science* 303, 102655.

Wells, C.W. (2024) Effects of essential oils on economically important characteristics of ruminant species: A comprehensive review. *Animal Nutrition* 16, 1–10.

Wilson, I.D. and Poole, C.F. (2023) Planar chromatography - current practice and future prospects. *Journal of Chromatography: B* 1214, 123553.

Wink, M. (2003) Evolution of secondary metabolites from an ecological and molecular phylogenetic perspective. *Phytochemistry* 64(1), 3–19.

Wink, M. (2008) Evolutionary advantage and molecular modes of action of multi-component mixtures used in phytomedicine. *Current Drug Metabolism* 9(10), 996–1009. DOI: 10.2174/138920008786927794.

Wink, M. (2015) Modes of action of herbal medicines and plant secondary metabolites. *Medicines (Basel, Switzerland)* 2(3), 251–286.

Wink, M. (2022) Current understanding of modes of action of multicomponent bioactive phytochemicals: Potential for nutraceuticals and antimicrobials. *Annual Review of Food Science Technology* 13, 337–359.

Wink, M. (2024) Duftstoffe - die Sprache der Pflanzen: Zur ökologischen Bedeutung von ätherischen Ölen. *Pharmakon* 6, 445–452.

Wink, M. and Schimmer, O. (2010) Molecular modes of action of defensive secondary metabolites. In: Wink, M. (ed.) *Functions and Biotechnology of Plant Secondary Metabolites*, 2nd edn. Wiley-Blackwell, Ames, Iowa, pp. 21–161.

Wojtunik-Kulesza, K.A. (2022) Toxicity of selected monoterpenes and essential oils rich in these compounds. *Molecules (Basel, Switzerland)* 27, 1716.

Wolfender, J.-L., Marti, G., Thomas, A. and Bertrand, S. (2015) Current approaches and challenges for the metabolite profiling of complex natural extracts. *Journal of Chromatography: A* 1382, 136–164.

World Health Organization (2003) WHO guidelines on good agricultural and collection practices [GACP] for medicinal plants. WHO, Geneva. Available at: https://www.who.int/publications/i/item/9241546271 (accessed 15 April 2025).

Xie, W., Weng, A. and Melzig, M.F. (2016) MicroRNAs as new bioactive components in medicinal plants. *Planta Medica* 82(13), 1153–1162.

Xue, F., Gu, Y., Wang, Y., Li, C. and Adhikari, B. (2019) Encapsulation of essential oil in emulsion based edible films prepared by soy protein isolate-gum acacia conjugates. *Food Hydrocolloids* 96, 178–189.

Yammine, J., Chihib, N.-E., Gharsallaoui, A., Ismail, A. and Karam, L. (2024) Advances in essential oils encapsulation: development, characterization and release mechanisms. *Polymer Bulletin* 81, 3837–3882.

Yang, Y., Xie, E., Du, Z., Peng, Z., Han, Z. *et al.* (2023) Detection of various microplastics in patients undergoing cardiac surgery. *Environmental Science & Technology* 57, 10911–10918.

Yin, Y. and Lee, S.-Y. (2020) Current view of ligand and lipid recognition by the menthol receptor TRPM8. *Trends in Biochemical Sciences* 45, 806–819.

Yoon, H.S., Moon, S.C., Kim, N.D., Park, B.S., Jeong, M.H. *et al.* (2000) Genistein induces apoptosis of RPE-J cells by opening mitochondrial PTP. *Biochemical and Biophysical Research Communications* 276(1), 151–156.

Yosri, N., Kamal, N., Mediani, A., AbouZid, S., Swillam, A. *et al.* (2024) Immunomodulatory activity and inhibitory effects of *Viscum album* on cancer cells, its safety profiles and recent nanotechnology development. *Planta Medica* 90(14), 1059–1079.

Yourick, J.J. and Bronaugh, R.L. (1997) Percutaneous absorption and metabolism of coumarin in human and rat skin. *Journal of Applied Toxicology* 17(3), 153–158.

Youssef, F.S., Mamatkhanova, M.A., Mamadalieva, N.Z., Zengin, G., Aripova, S.F. *et al.* (2020) Chemical profiling and discrimination of essential oils from six *Ferula* species using GC analyses coupled with chemometrics and evaluation of their antioxidant and enzyme inhibitory potential. *Antibiotics (Basel, Switzerland)* 9(8), 518.

Zani, F., Massimo, G., Benvenuti, S., Bianchi, A., Albasini, A. *et al.* (1991) Studies on the genotoxic properties of essential oils with *Bacillus subtilis* rec-assay and Salmonella/microsome reversion assay. *Planta Medica* 57(3), 237–241.

Zeng, Z., Zhang, S., Wang, H. and Piao, X. (2015) Essential oil and aromatic plants as feed additives in non-ruminant nutrition: A review. *Journal of Animal Science and Biotechnology* 6(1), 7.

Zhu, L., Li, R., Yang, K., Xu, F., Lin, C. *et al.* (2023) Quantifying health risks of plastisphere antibiotic resistome and deciphering driving mechanisms in an urbanizing watershed. *Water Research* 245, 120574.

Zimmermann, E. (2022) Aromatherapie für Pflege- und Heilberufe: *Kursbuch für Ausbildung und Praxis*, 7th edn. Karl F. Haug Verlag, Stuttgart, Germany.

www.ingramcontent.com/pod-product-compliance
Lightning Source LLC
Chambersburg PA
CBHW042314210326
41599CB00038B/7130